Green Politics and Neo-Liberalism

Also by Dave Toke

GREEN ENERGY

THE LOW COST OF THE PLANET

Green Politics and Neo-Liberalism

Dave Toke
Lecturer in Green Politics
Department of Political Science
and International Studies
University of Birmingham

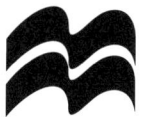

First published in Great Britain 2000 by
MACMILLAN PRESS LTD
Houndmills, Basingstoke, Hampshire RG21 6XS and London
Companies and representatives throughout the world

A catalogue record for this book is available from the British Library.

ISBN 0-333-77123-0

 First published in the United States of America 2000 by
ST. MARTIN'S PRESS, LLC,
Scholarly and Reference Division,
175 Fifth Avenue, New York, N.Y. 10010

ISBN 0-312-23588-7

Library of Congress Cataloging-in-Publication Data
Toke, Dave.
   Green politics and neo-liberalism / Dave Toke.
     p. cm.
   Includes bibliographical references and index.
   ISBN 0-312-23588-7 (cloth)
    1. Environmental policy. 2. Rational choice theory. 3. Discourse analysis. I.
Title.
   GE170 .T64 2000
   363.7'05—dc21

00-042241

© Dave Toke 2000

All rights reserved. No reproduction, copy or transmission of this publication may be made without written permission.

No paragraph of this publication may be reproduced, copied or transmitted save with written permission or in accordance with the provisions of the Copyright, Designs and Patents Act 1988, or under the terms of any licence permitting limited copying issued by the Copyright Licensing Agency, 90 Tottenham Court Road, London W1P 0LP.

Any person who does any unauthorised act in relation to this publication may be liable to criminal prosecution and civil claims for damages.

The author has asserted his right to be identified as the author of this work in accordance with the Copyright, Designs and Patents Act 1988.

This book is printed on paper suitable for recycling and made from fully managed and sustained forest sources.

10 9 8 7 6 5 4 3 2 1
09 08 07 06 05 04 03 02 01 00

Printed and bound in Great Britain by
Antony Rowe Ltd, Chippenham, Wiltshire

To Benjamin and James

Contents

List of Tables	x
Acknowledgements	xi
Introduction	1

1	Discourse, Power and Environmental Policy	5
	Foucault on power	9
	Discursive transformation	11
	The shifting self-interests of US CFC producers and consumers	15
	De-centred power is not enough	17
	Transport and power	20
	Alternative energy and power	26
	Truth and knowledge about US nuclear power	32
2	Rational Choice Theory and Environmental Policy	35
	The fall of classic rational choice	36
	RCT and 'soft' incentives	41
	Ostrom's institutional RCT	44
	Belief systems, norms and self-interest	47
	RCT and new international environmental problems	51
3	Science, Politics and Environmentalists	58
	In the shadow of the bomb	60
	From the scientisation of politics to the politicisation of science	62
	Scientists and environmental groups	64
	Saving the whale	66
	CFCs and ozone depletion	68
	Beyond positivism	70
	Blurred roles	72
4	Neo-liberalism and Green Politics	75
	Public choice and market theory	76
	The norms of public choice	78
	Middle classes and material self-interest	81
	Interpreting self-interest	82
	Can green politics reinterpret self-interest?	85

5 Health and Materialism	92
Inequalities, health and social malaise	94
Stress health and work	98
Status and the competitive society	102
Competition and consumption	105
The consequences of too much competition	106
The new discourse of stress at work	108
6 The Politics of Performance	113
Education, competition and performance	118
Does increasing competition in education work?	120
School league tables	123
Co-operative solutions	126
7 A Green Alternative	129
A culture of stress	129
Holistic approaches to society	132
Greens and socialism	134
Gorz's green socialism – a path to paradise?	136
The shape of future work	141
Reforming work	144
Paths to holism: eastern and western	147
Paths to 'downshifting'	149
Greens and urban regeneration[3]	154
8 Truth, Technology and Progress	159
Epistemology and ontology	161
The notion of 'species realism'	162
Human truths and environmental progress	166
Postmodernism and environmental practice	168
Technology and progress	173
Technology and risk	176
Three ages of control	180
9 Concluding Comments	184
Discourse and truth	184
Social green strategies	185
Reconstituting self-interest	187
Rational choice and self-interest	188
Correcting foucault	189

Notes	191
Bibliography	193
Index	205

List of Tables

5.1 GDP and life expectancy in selected countries (1993)   93
6.1 Attitudes to performance pay among UK Inland
      Revenue staff   117

Acknowledgements

I am greatly indebted to Dave Marsh in particular for providing me with a stimulating intellectual atmosphere and also plenty of ideas that have influenced this book. Matthew Paterson has been particularly useful in offering advice and comments. Members of the University of Birmingham's Department of Politics and International Affairs, such as Colin Hay and Peter Preston, have also been helpful and influential in various ways. Helpful comments have been received from others, including Robert Garner, Stuart McAnulla, Mike Davies, Horace Herring, Mike Cross and Dave Humphreys.

Dave Toke

Introduction

I have been inspired to write this book by a feeling that our modes of analysis as well as our ways of doing things are increasingly influenced by an individualistic, materialist and competitive creed that negates the values associated with green social and political ideas. It seems to me that if green politics is to be analysed, it cannot be done on the basis of accepting what passes for received notions of rationality. Hence my recourse to a discourse methodology and critiques of rational choice theory and neo-liberalism.

There are thus two key, interrelated themes in this book. The first is critiques of public choice theory and rational choice theory. The set of ideas, the discourse known as public choice theory says that the public good can best be achieved by the pursuit of individual, material self-interest and by minimising the role of the state. Public choice theory forms an important, perhaps the most important, part of the intellectual justification for the neo-liberal discourse that dominates the Anglo-Saxon world. Public choice theory is a subset of rational choice theory, which says that political outcomes can best be modelled by examining how individuals pursue their self-interest, which usually means their material self-interest. Compared to public choice theory, rational choice theory is a less politically tendentious method of analysis, but nevertheless its assumptions form part of public choice theory.

It is of course insufficient just to criticise. It is necessary to put forward a positive alternative. This is the second theme of the book. The critiques, and alternative modes of analysis, of public choice and rational choice theory are conducted at two levels: normative (what should happen), and analytical (how things happen). At a normative level I put forward green politics as an alternative to public choice theory and neo-liberalism. At the analytical level I put forward an alternative analysis based on discourse theory.

In the early part of the book I develop a method of analysis of environmental policy based on discourse theory, and I contrast this analytical approach to approaches based mainly on rational choice theory. In Chapter 1, I engage in an exposition and evaluation of Foucault's discourse approach to analysis, a mode that I have used to frame the discussion concerning neo-liberalism and green politics later on in the book. Discourse analysis forms a key part of the analytical alternative to modes of analysis that rest mainly on rational choice theory. I adapt the discourse method to take account of what I see as the shortcomings of Foucault's approach. This discussion is aided by practical examples drawn from transport policy and alternative energy policy.

Having laid out the basis of the discourse method as an approach to environmental issues, I move on, in Chapter 2, to evaluate rational choice theory as a tool of analysis. I analyse classical and modern rational choice theory, and discuss whether it is possible for modern rational choice theory to escape from the criticisms made of classical rational choice and public choice theory. Chapter 3 seeks to chart the development of the environmental discourse and its interaction with science. I look at the role of agents of this discourse – that is, scientists and environmental pressure groups.

The fourth chapter signals the start of an attempt to use the analytical critiques and uses of discourse theory and rational choice theory that I have developed in an expanded political critique of, and alternative to, the most prevalent discourses associated with rational choice theory. These are, as already noted, public choice theory and, especially, neo-liberalism. I outline the nature of the neo-liberal discourse and look at the possibilities for green politics to reinterpret self-interest.

In the fifth chapter I discuss health and materialism with an emphasis on stress. Here I demonstrate that the untrammelled pursuit of individual material self-interest is neither necessarily in the interest of the individual nor of society as a whole. I do this by examining how the increasing obsession with competition is damaging the health of both individuals and societies. In the sixth chapter I describe some techniques associated with neo-liberalism, such as performance-related pay and increased competition in education, and I discuss some of their negative features. I sketch out some green alternative ideas in Chapter 7.

Chapter 8 continues the process of adaptation of the discourse method in the context of trying to develop a theory of knowledge that can act as a basis for charting environmental progress and analysing the role of technology. Finally, in Chapter 9, I make some concluding comments.

1
Discourse, Power and Environmental Policy

The purpose of this chapter is twofold: first, to discuss the advantages and limitations of Foucault's ideas on discourse and power; and second, to illustrate such advantages and limitations in the context of an analysis of some environmental issues, notably including ozone depletion, transport policy and alternative energy policy in the United Kingdom.

This is certainly not the first time that discourse analysis has been used to analyse environmental issues and policy. Martin Hajer (1997) has penned an incisive and painstaking effort to combine discourse theory, the theory of ecological modernisation and analysis of the acid rain policy field in the United Kingdom and the Netherlands. It is entitled *The Politics of Environmental Discourse*, and it charts how 'discourse coalitions' operating with an ecological modernisation perspective challenged the 'traditional pragmatic' discourse on pollution control in the United Kingdom with only partial success. The ecological modernisers were more successful in the Netherlands in achieving comprehensive discursive change without, however, achieving full implementation of all the practices that the ecological modernisers associated with their discourse.

Karen Litfin's work on the ozone depletion story, *Ozone Discourses* (1994), focuses much more than I do on international relations theory. However, what she says about the role of science in the run-up to the conclusion of agreements to protect the ozone layer is revealing. It was not, or at least not just, science that paved the way for the Montreal Protocol. She analyses the key importance of the physically non-existent, but metaphorically powerful, 'hole' in the ozone layer.

A third work is John Dryzek's *Politics of the Earth* (1997), which adopts an idiosyncratic, although eminently serviceable, approach to the discursive method of analysis. Dryzek explicitly rejects the mainstream use of the discourse technique to analyse how particular discourses become dominant. Rather, he uses the technique as a means of describing and analysing different political approaches to environmental problems. In contrast, my main approach is to establish how discourses are established and/or become dominant. I talk about discourses being dominant in a particular sphere. Generally this means a dominant discourse in government policy.

Finally, there is *Discourses of the Environment*, a work edited by Eric Darier (1999), which is mainly concerned with linking Foucault's general theories to discussions of the environment. That such a task is important is undeniable, although none of the authors look, at specific green issues. It is understandable that various authors dealing with critical theory and the environment have ignored Foucault, given that Foucault said practically nothing about ecology, apart from commenting that ecologists pursued a different set of truths from those of established science (1999: 4). This is perhaps unsurprising given his apparent dislike of the countryside! On one trip to visit churches and museums in the Alps he is said to have ostentatiously turned his back when his friends pointed out to him magnificent views of landscape (1999: 6). However, I am sure Foucault would not have demurred from the potentially great purchase in using discourse analysis in a green political context.

I now want to describe and analyse Foucault's discourse theory in a way that is relevant to a study of environmental policy and green politics, although this involves covering what are regarded as two distinct phases in Foucault's development of his method. The first phase concerns the elaboration and illustration of the emergence and changes in discourse in various scientific, academic and professional disciplines. The second phase was focused more on analysis of power. Common to both phases is Foucault's demonstration, in a series of studies on different disciplines, how the disciplines, which involve discourses, had not developed as part of some process of progressive discovery of truths. Rather, they had developed through an erratic process of changing practices and ideas connected to historical contingencies and social requirements. Under Foucault's 'perspectivist' notion of truth it is not that there is no such thing as truth, but that claims to truth cannot be absolute claims, merely claims that are context- and discourse-specific.

Foucault charts how each discourse, specific to a particular discipline and historical period, has its own set of truths. For example, the notion of madness developed away from the medieval, religiously inspired belief that madness was morally deviant, towards seeing it as being a disease which needed to be treated. The context of this change was the development of industry, the emergence of capitalist economic interests and the associated development of the philosophy of reason (Foucault 1973, cited by Shapiro 1981: 142). There is a common theme in that Foucault seems to accept the key influence of the emergence both of industrialism and of the bourgeoisie, although this should not be confused with an acceptance of Marxism.

Shapiro (1981) illustrates the key importance of discursive dominance (sometimes called 'discursive hegemony', a term which I avoid because of its Marxist overtones) in deciding between different claims to truth. He discusses the contrast between the Marxist theory of value and the liberal economic discourse in explaining the same domain. The Marxist discourse is an asset to workers since value is attributed to the workers' labour. However, the liberal economic discourse (or neo-liberal discourse, as I discuss it) does not recognise the workers as a class with legitimate bargaining rights, thus reducing and altering power relationships to their disadvantage. The displacement of the importance of the labour theory of value, as well as the Keynesian theory of demand management (which was a discursive asset to the state), coincided, as I observe in Chapter 4, with the enlargement of the middle classes. However, it could be argued that the discourse on stress developed later in this book has the potential to recharge the assets of workers because of two emphases; first, the emphasis on the holistic notion of the health of the community being influenced by the degree of co-operation and equality at the workplace; and second, the emphasis on the idea that the health of the individual is influenced by their working conditions.

The contrast between neo-liberal and Marxist discourses should not lead us to imagine that pre-conceived discourses 'compete' with one another. Foucault was keen to point out that discourses consist of elements which could be recombined, omitted, added and generally organised so as to produce dramatic alterations in the significance of the discourse. Hall, for example, analysed how Thatcherism represented, in Howarth's words,

> the articulation of two seemingly contradictory sets of theories...neo liberal economics (the social market, self interest, monetarism, competitive individualism) and the older philosophies of organic Conservatism (nation, family, duty, authority, standards, traditionalism).
>
> (Howarth 1995: 125)

Some may use the term 'ideology' in place of discourse, but this can carry the implication that discourses are, on the one hand, somehow integrated wholes and, on the other hand, projected by class formations in the Marxist sense. As neither sense is implied here, the term 'ideology' is not used.

In his earlier period Foucault tended to concentrate on the content, nature and rules governing discourse formation and change rather than charting the circumstances in which they changed. In *The Archaeology of Knowledge* (1994) he focused on the existence of a series of rules which governed what he called a discursive formation; for instance, involving medicine, economics, psychiatry or natural history. He said that 'discourse can be defined as the group of statements that belong to a single system of formation' (Foucault 1994: 107). Foucault talked about discursive formations having objects (items that emerge from a discourse), concepts, 'enunciative modalities' and theories or strategies. In psychiatry, an object might be a person classified as mad. A concept in natural history might be that of a species. An enunciative modality involves those persons and institutions responsible for authoritative articulation and interpretation of the discourse, such as doctors or hospitals in the case of medical discourse. A theme or strategy – for example, the liberal or neo-classical economic discourse as applied to economics – involves a theory which links various concepts and the behaviour of objects. According to this 'archaeological' method, changes in a particular discourse could be observed by noting the changes in these categories and the rules that govern the discursive formulation. For example, as Davidson explains in the case of sexuality:

> in the mid-nineteenth century, a mutation in the rules for the production of discourse first made it possible to speak about sexuality, and not merely about sex. These same rules permitted new talk about diseases of sexuality, allowing doctors to isolate these diseases as distinct morbid entities and bringing about an unprecedented discourse on perversion.
> (Davidson 1986: 226)

In fact, by the time Foucault wrote his volumes on *The History of Sexuality* (1981), he had moved on to focus on a discussion of power. His earlier 'archaeology' was associated with a structuralist mode of thought that was becoming unfashionable. Why, for example, should Foucault's own typology of the structure of discursive formations be the only possible typology and why should the typology apply with equal relevance to all discourses? There is a further problem when one comes to use his archaeology to analyse ecology, in that ecology does not exist as a discursive formation in the terms discussed by Foucault. A multitude of differing understandings of the term 'ecology' exist, and such understandings even differ in the apparently proximate fields of botany and zoology (Remmert 1980: 2). As Hajer observes, 'a discussion of a typical environmental problem involves many different discourses' (1995: 45), and Hajer mentions several when he introduces the topic of acid rain. Indeed, Foucault himself admits to 'abusing' the term 'discourse'. The use of the term to describe groups of statements that fall far short of Foucault's rigorous specification of those that comprise a discursive formation is widespread, and I certainly use the term in this attenuated form when I discuss practical examples of discourses later on.

Nevertheless, the archaeological method can have important analytical uses with regard to environmental issues if one looks at the rules of formation discussed by Foucault not as the uncovering of some preordained structure, but as one possible framework (among others that could conceivably be devised) for analysing a particular policy area. Use of the method may help us establish a footing, or at least help us clarify our approach so that the development of policy in an area can be examined. We can structure our discussion by investigating, in the relevant field of interest, how and when objects emerge, how concepts are formed and developed, who and where is the location for making authoritative statements containing truths. We can also investigate the often conflicting strategies that are promoted to analyse or deal with a particular problem.

Although Foucault's earlier writings were dominated by a study of the emergence and nature of discourses, an equally important lesson taken from Foucault in this book is the lesson on power.

Foucault on power

Hajer regards Foucault's later 'genealogical' emphasis on power as being more central to his methodology than the previous emphasis on the archaeology of discourse. Stung by Marxist criticisms that the discourses he was studying were reflections of changing eddies of class power, Foucault sought to study the nature of power, in particular what he called the micro-physics of power. He saw that the nature of power relations emerges from a disaggregated and complex process of fusions of interests and ideas in very specific situations. The ideas act to constitute the interests, which in turn generate ideas that interact with, and constitute, new interests. Consequently Foucault felt able to chart how discourses changed, how new dominant discourses emerged, rather, in the earlier archaeological period of his writings, than merely describing the different forms of truth that rose and fell in an apparently arbitrary manner. As Poster puts it, in contradistinction to Marxist formulations: 'Discourses are important because they reveal the play of power in a given situation. They are not "ideological representations" of class positions but acts of power shaping actively the lives of the populace' (Poster 1984: 130).

Whereas in the earlier phase of Foucault's studies power seemed to be associated with linguistic formations and their use, in the genealogical phase power arose from an interaction between discourse and practice and involved, crucially, the constitution of the subject; that is, people's identities:

> One has to dispense with the constituent subject...to arrive at an analysis which can account for the constitution of the subject within a historical framework. And this is what I would call genealogy, that is a form of history which can account for the constitution of knowledges, discourses, domains of objects etc. without having to make reference to a subject which is either transcendental in relation to the field of events or runs in its empty sameness throughout the course of history.
>
> (Foucault 1980: 117)

A key feature of Foucault's theory of power which needs to be borne in mind is his notion that power is a de-centred conception. Power is not possessed by specific agents: it is a field of relations. When discussing the development of modes of discipline and punishment Foucault commented: 'The power exercised on the body is conceived not as a property...one should decipher in it a network of relations, constantly in tension, in activity, rather than a privilege that one might possess' (Foucault 1977: 26).

Later on in the same passage Foucault discusses how power and knowledge are intimately related:

> We should abandon a whole tradition that allows us to imagine that knowledge can exist only where the power relations are suspended. ...We should admit rather that power produces knowledge...that power and knowledge directly imply one another; that there is no power relation without the correlative constitution of a field of knowledge, nor any knowledge that does not presuppose and constitute at the same time constitute power relations. ...It is not the activity of the subject of knowledge that produces a corpus of knowledge, but power-knowledge, the processes and struggles

that traverse it ... that determines the forms and possible domains of knowledge.

(1977: 27–8)

We can see from this passage the importance of studying the development and implementation of dominant discourses. For power, according to Foucault, is derived from such discourses, not the agents themselves. For example, the power of employers to hire people on short-term contracts or implement systems of performance-related pay is dependent, not on the power of the employer as an agent, but on the existence of a dominant discourse that generates and is associated with a body of knowledge that confers legitimacy on such practices. This field of power relations also limits the ability of unions to resist such policies. It will only be when the contemporary version of the neo-liberal discourse ceases to be dominant that there can be effective resistance to such policies.

Another aspect of Foucault's 'micro-physics of power' are new ways of moulding the subject and exercising power over the body. Foucault charts how there has been an increase in the deployment of power through knowledge in society generally since the beginning of the Industrial Revolution. This form of power, a growth in political technology, involves, for example, increased scrutiny and knowledge of the actions and subjectivities (the nature of souls) of prisoners through ensuring constant observation of their activities. This type of power is said to have replaced the repressive authority that maintained order in feudal times. However, these developments in machine technology as well as political and social technology have increased the scrutiny of rulers as well as ruled. Presidents Nixon and Clinton found that modern devices like tape-recorders and DNA testing allowed their detractors to hold them to account to a degree that would not have been possible in times before the development of such technologies.

Let us now move towards developing techniques for analysing discursive transformation in the environmental sphere through a continuing discussion of policy on ozone depletion and a comparison with the discursive development of policy on global warming.

Discursive transformation

In order to fashion tools for analysing discursive transformation we need to trace how environmental discourses develop, how such discourses become dominant, and how changed policies and practices occur. The latter stage is usually associated with a discourse becoming 'embedded' in social culture, a term I borrow from Christoff (1996). For example, we can see that the 'ozone depletion' discourse emerged as an alternative discourse in the early 1970s, initially in response to fears about the dangers of ozone depletion being caused by emissions from supersonic aircraft. This then became focused on the impact of halocarbons, chiefly chlorofluorocarbons (CFCs), on stratospheric ozone concentrations. Environmental discourses contain elements suggesting that a particular human-induced activity or series of activities (for example, CFC use) is causing potential or actual environmental problems (in this case ozone depletion) and that certain actions should be taken to stop this happening (such as phasing out CFC use). This halocarbon–ozone depletion discourse became dominant in several western states the early 1980s after the European Union (EU) and other countries agreed to cut CFC use, and the Vienna Convention was signed in March 1985 with the aim of protecting the ozone layer. However, these measures would do little to curb the problem. There was relatively little public awareness of the issue, until, a couple of months after the signing of the Vienna Convention, a British Antarctic Survey revealed unexpected declines in ozone concentration over the South Pole (Litfin 1994: 73–7). This received tremendous publicity and the Montreal Protocol and London Protocols, which called for CFCs to be phased out, followed. The agreement of the Montreal Protocol in 1987 coincided with consumer boycotts of and legal actions against uses of CFCs. The US CFC industry changed its tack, about which more will be said later in the chapter, and the United States began to take the lead in calling for curbs on CFC production. This led to the adoption of the Montreal Protocol. This was the 'embedding' phase of the discursive development of the issue. This induced industry to find solutions that proved, in the end, to be much cheaper than originally anticipated, and achievable mainly on the basis of changing psychological attitudes to adapt existing technology (Litfin 1994: 118, 157).

One can analyse the global warming issue using this model. The global warming discourse has become dominant at the governmental level in many western states in that governments have agreed to aim for small cuts in carbon dioxide emissions. However, the problem still seems to be difficult to solve because the global warming discourse has not yet become 'embedded' in the minds of the public. Dunlap (1998), reporting on the findings of a comparative study in attitudes to global warming, said:

> Responses to this open ended question [concerning the causes of global warming] suggest that throughout the six nations citizens

have a poor understanding of the causes of global warming. As found in numerous in-depth studies of small, convenience samples, lay publics are most likely to see global warming as being caused by air pollution or ozone depletion, and in no nation do as many as a third volunteer (in their first response) fossil fuel use or deforestation as the main causes.

(1998: 482)

In the United States, 29 per cent of those asked to volunteer a cause for global warming cited 'pollution', 25 per cent blamed 'CFCs or ozone' and only 22 per cent blamed 'fossil fuel use'. When asked to rate various named causes of global warming, 38 per cent of the respondents named nuclear power as a cause of global warming compared to only 29 per cent who rated refrigerators and air conditioners to be causes of global warming. Women were significantly more likely to see nuclear power as a cause of global warming than men, a phenomenon that was ascribed, by the authors, to women's stronger anti-nuclear orientation.

Given the lack of awareness of global warming it is likely that awareness of, and belief in the importance of, solutions to the problem are likely to be even more threadbare. This is borne out by a survey of consumer attitudes to buying refrigerators. The survey found that only 2 per cent of UK consumers regard environmental friendliness as being the most important factor in buying a fridge. However, according to the same study, on average, relatively energy-efficient fridges are no more expensive to buy than relatively energy-inefficient fridges, even though the efficient models are much cheaper to run (Boardman 1997: 25).

Lucas (1985: 278–9) discusses the psychological barriers in industry and commerce to taking advantage of the great range of energy-efficiency opportunities that exist. Chakravorty *et al.* (1997) attest to the possibilities for solving the global warming problem through declining prices for electricity from solar power. Certainly President Clinton has promoted a subsidy programme for the growing solar electricity industry.

Lovins and Lovins (1998) have argued that consistent and continuing energy efficiency savings can be achieved when energy efficiency is made a major priority and where learning becomes embedded in the local culture:

In 1981, Dow Chemicals' 2400-worker Louisiana division started prospecting for overlooked savings. Engineer Ken Nelson set up a shop-floor-level contest for energy saving ideas. Proposals had to offer at least 50 per cent annual return on investment (ROI). The first year's projects averaged 173 per cent ROI. Startled at this unexpected bounty, though expecting it to peter out quickly, Nelson persevered. The next year 32 projects averaged 340 per cent ROI. Twelve years and almost 900 implemented projects later the workers had averaged...204 per cent audited ROI. In the later years, the returns and the savings were getting bigger, because the engineers were learning faster than they were exhausting the 'negawatt' [negative megawatt] resource. ...By 1993, the whole suite of projects was paying Dow shareholders $110 million every year.

(1998: 2)

The point that needs to be made here is that, in order for people to take advantage of even existing technological opportunities they need to 'know' that taking action to counter global warming is important, or at least, as in the case cited above, people need to 'know' that there are other benefits including considerable financial savings. If they do, if the global warming discourse becomes embedded, then the solutions may be a great deal cheaper than they appear to be at the moment. Engineers, designers and consumers would come to regard energy efficiency as a high priority, and, as we shall see in the next section, this knowledge becomes power, which transforms technological structures.

This model of discursive development of environmental issues should not be interpreted in a structuralist manner, merely as a means of comparing similarities that might exist between different environmental issues. For example, in the case of genetically modified organisms (GMOs) and food, the anti-GMO discourse seems to have become embedded in some European states before it achieved the status of a dominant discourse at governmental level. In this case the discourse is effectively being implemented through consumer pressure on food retailers. In the United States anti-GMO feeling has taken much longer to be embedded, but even there, there were, at the time of writing, growing pressures to legislate for genetically modified food to be labelled. In the autumn of 1999, Monsanto, the leading company producing genetically modified food, fearing that the American market would follow the European market, signalled a desire to talk to environmentalists about alternative organic farming methods (Lean 1999).

We can see that the study of dominant discourses is crucial if we are going to understand the factors which act to constitute actors' perceptions of their own self-interest – that is, constitute the nature of the subjects in a particular policy field. The nature of the consumers and producers, as subjects, is reconstituted by the changing natures of the dominant discourses, the changing truth regimes, which affect how particular products and potential products are seen. We can see this process in operation in the case of US CFC producers.

The shifting self-interests of US CFC producers and consumers

Legislation limiting CFC use for aerosols was passed in the United States in 1977. However, the limitations on CFC production were not very successful even in the United States, the only country where any controls had been instituted. Despite the dominance of the halocarbon–ozone depletion discourse among relevant scientists and governmental environmental agencies, the chemical industry remained in a stage of denial in defence of its interests. In the dying days of the Carter Administration the Environmental Protection Agency (EPA) drafted a rule-making proposal for the regulation of CFCs. In order to combat what CFC producers then saw as a threat, some 500 CFC users and producers formed an 'Alliance for a Responsible CFC Policy' (Litfin 1994: 70). Successful lobbying of the Reagan Administration to abandon any plans to regulate CFC production led to the abandonment of efforts, by Du Pont, to research into a chemical substitute for CFCs. Right up until 1986 Du Pont claimed there were no credible substitutes for CFCs. However, attitudes began to change dramatically following the publication in May 1985 of results from the British Antarctic Survey research project, which presented clear evidence of dramatic declines in ozone concentrations over the Antarctic. These results were pictured as showing a 'hole' in the ozone layer. In fact, there is no such thing as an ozone layer in the commonly accepted notion of the term, never mind a hole. It was just that concentrations of stratospheric ozone had greatly declined. The ozone hole was, as Litfin has said, a metaphor that shifted discourse from description from prescription. 'The ozone hole created a sense of crisis that was conducive to the precautionary approach eventually sanctioned in the Montreal Protocol' (in 1987) (Litfin 1994: 97).

The effect of the discovery of what became known as the hole in the ozone layer was to change people's sense of self-interest. Hajer comments: 'Interests cannot be taken as given a priori but are constituted through discourse' (1997: 51). Consumers now saw their interests, as purchasers of aerosols that might contain CFCs, not merely as people seeking a deodorant that matched their self-image, but also in terms of whether they were agents of environmental protection or damage. One could use economists' language and say that the consumers' preference structures changed, but in Foucauldian terms this is a symptom of a subtle change in consumer identities and, flowing from this, a change in consumers' perceived self-interest. We can see that the change in consumer behaviour cannot be studied adequately through an atomistic study of individual preferences since the changing consumer behaviour is brought about by discursive change. This discursive change is not solely concerned with one issue but involves the emergence of a socially conscious, 'green' consumer-subject who is often likely to be sensitive to environmental discourses in general.

The discursively induced changed notion of self-interest infected the chemical companies just as much as anybody else. The chemical producers now saw their interests in responding to public fears that their products might be damaging the ecosphere. In 1987, legislation was filed in Congress calling for cuts in CFC use. Du Pont, followed by other CFC producers, rapidly shifted their position – first, to a recognition that substitutes to CFCs existed (which they could manufacture), and then, in 1988, to support for more stringent controls on CFC production. As one analyst of international ozone politics puts it, 'Du Pont, the world's largest CFC producer, discovered that international CFC regulation was in its interests. Du Pont was prescient enough to recognise that strict regulations on CFCs and halons could result a new market for substitutes if the appropriate market conditions were in place' (Seaver 1997: 95). Du Pont's reconstituted sense of self-interest, which occurred soon after the 'embedded' discursive change, also reconstituted the perceived interests and very identities of many consumers of aerosol and other CFC products who spurned products that still used CFCs. The power that forced the CFC producers' hands did not arise from the actions of specific agents, whether scientists or environmentalists, but from changes in dominant discursive structures. In other words, we are talking about the effect of de-centred notions of power. This type of power is a central concept of Foucault's.

So far the discussion has been limited to explaining Foucault's modes of analysis, giving some examples, and extending the analysis in the field of environmental policy. Now I want to look at the possibilities for amending some aspects, particularly as regards his notion of de-centred power. Although Foucault's de-centred notions of power may be enlightening in various ways, they are still limited in certain respects, especially by Foucault's refusal to consider that power should not only be considered as a field of relations, but also as a property of particular actors.

De-centred power is not enough

Much of the reasoning for Foucault's de-centred conception of power seems to stem from his theory of knowledge, which is derived from Nietszche. This perspectivist viewpoint involves the assumption that complete truth can never be told and that precise causes cannot be attributed to effects. Foucault implies that specific attributions of power cannot be ascribed to precise agents, even if the general influence of agents in moulding structures can be recognised. He even refuses to define what power is, as opposed to saying what it is not.

While the dubious nature of attributing precise cause-and-effect relationships should be recognised, this does not necessarily rule out the drawing of some general conclusions about whether particular groups have exerted some degree of influence in a particular situation, even if the exact degree of influence cannot be determined in isolation from a range of existing and changing dominant discourses and participating agents. Foucault's refusal to attribute power to precise actors does not help answer certain common practical questions both in political science and public debate. In the environmental field these include questions such as 'How powerful are environmental groups?' and 'Are insider rather than outsider pressure-group strategies more effective?' The demands of academics for answers may be brushed aside at some cost, but these questions are asked in a very practical sense by activists, journalists and policy analysts who will, as a matter of need, answer it by themselves if political scientists do not offer help.

Grant (1995) and other pressure-group specialists, such as Maloney *et al.* (1994), tend to suggest that outsider groups, which put the bulk of their energy into influencing public attitudes, are much less influential than insider groups, which concentrate on lobbying civil servants or ministers. Grant suggests that 'insider groups have a better chance of influencing government policy' and that 'in the longer run most groups tend to veer towards an insider strategy because of the gains it offers' (1995: 19–20). It has been claimed that groups such as Greenpeace, which deliberately choose the outsider strategy do so essentially as a membership recruitment strategy 'even though this is likely to prevent success in the policy area' (Maloney *et al.* 1994: 32). On the other hand, analysts such as Doherty (1998) argue that, in the case of the United Kingdom's road-building controversy: 'without the protests of direct action groups the issue would not have remained at the centre of national political debate' (1998: 383). The implication here is that outsider tactics can be very effective. Who is right in this debate?

It seems that Foucauldian notions of power cannot help answer such questions. However, if we have a notion of power whereby power can be ascribed to individual groups as well as to fields of relations, it may be possible to argue that outsider groups can be said to wield considerable power in a specific situation. They could be said to wield power when they are able to bring about a change in the dominant discourse, even though they may lose specific 'battles'.

In fact, there are cogent arguments to suggest that Foucault's formulations themselves contain the seeds for the attribution of power in a centred as well as a de-centred way. Even if we look at the 'structuralist' *Archaeology of Knowledge* we can see this. Foucault comments that 'If, in clinical discourse, the doctor is in turn the sovereign, direct questioner, the observing eye, the touching finger, the organ that deciphers signs, the point at which previously formulated descriptions are integrated, the laboratory technician, it is because a whole group of relations is involved' (1994: 53).

Foucault, in describing the role of doctors, was describing their role quite clearly within a group of relations which determines what the doctor says in diagnoses, prognoses and so on. Nevertheless, there is at least the suspicion of an agency role, as much as Foucault may wrap it up in his networks of relations, and as much as he may claim the role to be nominal in character. In the case of environmental discourses, the role of the environmental group or scientist(s) is very active both in forming/transforming the discourse and also in helping to propel an initially marginal discourse into a position of dominance. This is as compared to the passive, nominal role of a doctor as merely representing the medical discourse. What I am saying is that by stressing and enlarging upon the 'suspected' agency aspect of the enunciative modality formulation, it is possible to present an interpretation of power, adapted, albeit with changes, from Foucault, that is both centred and de-centred, that not only takes account of the power of knowledge and of actions associated with certain discourses but also the role of specific agents in achieving discursive transformation.

Foucauldian purists may protest about this adaptation of Foucauldian notions of power on epistemological grounds, among others. However, philosophers like Wittgenstein, whose later work runs parallel to that of postmodernists and proponents of the discourse approach in terms of their pragmatic approach to the notion of truth, say that the meaning of language, the difference between sense and nonsense, is guided by its use. If we employ language because it is useful, and not because it is justified by a theory of language, can we not use terms in political analysis according to their utility in answering commonly asked questions? This is as opposed to their theoretical correctness, especially given that philosophical theories are devices that are generally deprecated by philosophers like Rorty and Foucault. In the case of power as a concept in general, can we not adapt our model of power to fit the needs of political analysis and, most pertinently, the demands placed upon it by people who can make use of such modelling? Thus we may opt to study the relative effectiveness of agents such as environmental groups operating with different strategies and, as is done later in this work, we may endeavour to compare the roles of environmental groups and epistemic communities (groups of experts) in the manufacture and dissemination of knowledge. We may do these things because they serve a practical purpose, not because they represent absolute truths.

In proposing a centred and de-centred notion of power, it may be convenient to adapt some formulations explained by Hay (1997). Hay has surveyed the mainstream conventional literature on power focusing on the so-called 'three faces' of power. The first face of power is based on Dahl's (1957) definition of power as being exerted when actor A makes B do something that B would not otherwise have done. Bachrach and Baratz (1962) point to a second face of power, which involves the ability to set decision-making agendas to exclude certain options. Finally, Lukes (1974) identifies a third face of power, which highlights how people's conceptions of their own self-interest are manipulated so that they do not demand action that will fulfil their 'objective' interests. Hay criticises Lukes, correctly, for the patronising idea that an outside analyst can say what someone's objective interests are. Instead, Hay proposes a conception of power that retains Dahl's direct notion of power involving A influencing B's actions, but supplements this with the indirect power of an actor to influence the context in which others operate (1997: 50).

Although Hay's notion of power is still focused on the actor, his version is crucially different in that it involves influencing the context. However, Hay still fails to acknowledge the influence of discursive structures – in particular, dominant or emerging discourses. Discussing the interaction of structure and agency, he comments that:

> Power is a question of agency, of influencing or 'having an effect' upon the structures which set contexts and define the range of possibilities of others. This suggests the need for a relational conception of both structure and agency: One person's agency is another person's structure. Attributing agency is therefore attributing power (both causal and actual).
>
> (Hay 1995: 191)

But, if both structures and agents influence the actions of others, if one person's agency is another person's structure and if attributing agency is attributing power, then it follows that attributing structure, through the study of discourse, is also attributing power. As such we need a further aspect of power to supplement Hay's direct, Dahl-type power of one agent over another and his power of an agent to influence the context in which others operate. This supplementary, third type of power must be a power inherent in structures that manifest themselves in the form of discourses. This third power is power/knowledge, which we recognise in Foucault's analyses.

Some, like Dowding (1991), reject the notion that structures as well as individuals can have power. He says that the notion is redundant and conceptually misleading. In particular, he points out that someone's position in the social structure limits an actor's choice but it does not determine it (1991: 8–9). This latter statement is true enough, but then one can apply the same sort of criticism to individual ascriptions of power. Dowding concedes that individual power is in part determined by structure, so if power is only partly determined by choice, why is individually ascribed power not misleading in the same sense as structural power? I would argue that discourse is a special type of structure. It shapes and transmits beliefs and norms and therefore it helps political analysis if one understands discourse as a power relation. This becomes clearer when one considers Dowding's notion of 'systematic luck' (1991: 137–8, 152–7) as an explanation as to why farmers and capitalists have a tendency to win political favours. Dowding says that luck flows from a favourable political background. Maybe, but that favourable background does not consist of unmediated events that favour certain groups; it consists of a discourse which carries sets of beliefs which interpret the events. This discourse influences policy outcomes. The 'luck' pours out of a dominant discourse, and so it is meaningful to ascribe power to that discourse. In this sense it is Dowding's notion of systematic luck that is redundant, not the idea of power/knowledge associated with discourses.

I shall endeavour to illustrate the utility of this hybrid mixture of Foucauldian and more conventional conceptions of power in two case studies. The first deals with road-building and transport policy in the United Kingdom.

Transport and power

In the United Kingdom, transport has been a battleground between environmentalists and the road-building lobby since the beginning of the 1970s. Although a peace deal involving a less rapid building programme and a public inquiry system was established in the late 1970s, the abandonment of this deal was signalled by the publication of the Government *White Paper, Roads for Prosperity*, in 1989 (Department of Transport 1989). Launched in the context of bullish projections about traffic growth and a stated Thatcherite preference for the 'Great Car Economy', the *White Paper* was mainly concerned with outlining and justifying a new, ambitious road-building programme. Only one short paragraph out of 45 was devoted specifically to the environment. The *White Paper* states:

> Road congestion is bad for the economy. It imposes high costs on industry and other road users by wasting time, delaying deliveries and reducing reliability. ... There is no way of making accurate overall estimates [of the costs of road congestion], but it is clear that the costs are very high.
>
> (1989: para. 6)

Later on, the *White Paper* comments:

> The Government's conclusion is that the main way to deal with growing and forecast inter-urban road congestion is by widening existing roads and building new roads in a greatly expanded road programme. The scale of the problem is such that it can be relieved only by a step-change in both the size and the composition of the programme.
>
> (1989: para. 16)

The programme would

> improve the inter-urban motorway and trunk road network by reducing journey times and increasing the reliability of road travel. It is a vital further boost for British industry. The measures proposed will provide the means to improve the country's geography.
>
> (1989: para. 3)

It would thus help both prosperous and less favoured industries cope with, and achieve, growth.

An alternative discourse was put forward by an alliance of environmental interests which saw roads, not as a solution but as a problem 'which dislocated communities, despoiled the countryside and, by encouraging traffic, brought increased atmospheric pollution' (Dudley and Richardson 1996: 579). This discourse was supported by a wide coalition of interests ranging from Earth First!, a radical group which prioritised direct action, through Friends of the Earth (FOE), who supported non-violent direction action and who publicised and researched the issues and actions, to the relatively conservative Council for the Protection of Rural England (CPRE), which have much support from what is now called 'middle England'. Initially, the Government's strategy seemed to be winning through, despite the opposition, and road-building increased. However, by 1995 there had been a sharp change in both the tone and content of Government policy. The Minister for Transport, Brian Mawhinney, expressed doubts about the 'Great Car Economy' and the first of a series of large cutbacks in the road-building programme was announced. Analysis of figures released by the Department of the Environment, Transport and the Regions (DETR 1998a) reveals that in 1996–97 expenditure on all trunk roads had fallen by around 20 per cent compared to 1993–94, and by 1997–98 by nearly 30 per cent compared to 1993–94. Given that these figures for road spending include road maintenance, this decline means that the bulk of planned new road schemes were scrapped.

This change in policy was accompanied by considerable change in the dominant discourse at the Department of Transport. In 1996, in the *White Paper Transport, the Way Forward*, the transport debate was reconstructed so that increasing road traffic was conceived as a problem which was feared would lead to 'increased congestion and increased environmental pollution' (Department of Transport 1996: 121), rather than as a necessary consequence of economic progress which had to be accommodated by road building. Unlike the 1989 *White Paper*, there was no longer an uncritical acceptance of the economic value of roads. In 1994, the Standing Advisory Committee on Trunk Road Assessment (SACTRA) supported claims that new roads generated traffic growth and was authorised to undertake further study on the links between traffic growth and economic growth (Department of Transport 1996: 27). The 'roads' discourse became a 'transport policy' discourse, with investment in public transport and other strategies seen as a bigger priority than spending on road building. In future the roads budget would 'focus, increasingly on maintaining and managing the capacity of existing roads, and selective improvements through new construction, such as providing much needed bypasses and removing bottlenecks' (Department of Transport 1996: 65).

The incoming Labour Government's policy, described in the *White Paper A New Deal for Transport* (Department of Environment, Transport and the Regions 1998b) as an 'integrated transport policy', utilised a similar discourse, except that the focus on road building was further reduced.

The sentence '[t]he priority will be maintaining existing roads rather than building new ones and better management of the road network to improve reliability' (1998b: 3) was similar to the key sentence in the 1996 *White Paper* except that the reference to bypasses was omitted. This was emphasised by Prescott's foreword to the *White Paper*, when he said:

> The previous Government's Green Paper (*Transport the Way Forward*) paved the way with recognition that we needed to improve public transport and reduce dependence on the car. Businesses, unions, environmental organisations and individuals throughout Britain share that analysis. The White Paper builds on that foundation. …In its Green Paper the previous Government recognised that we could not go on as before, building more and more new roads to accommodate the growth in car traffic. With our new obligations to meet targets on climate change, the need for a new approach is urgent.
>
> (1998b: 3)

We can see that large elements of the environmentalist discourse on road building were incorporated into what became the dominant discourse on transport policy. In this sense the influence of environmentalism in this policy area was far-reaching.

These changes in the dominant discourse should be seen as a consequence of changes in power relations resulting from the manufacture and dissemination of new knowledge at events such as anti-roads protests and subsequent acceptance of this knowledge by key sections of middle-class opinion. There had been a string of strong objections to proposed roads, some successful at the planning stage, in the late 1980s and early 1990s (Toke 1997). However, the so-called 'Battle of Twyford Down' over the M3 extension, which ran from 1992 to 1993, together with other campaigns in 1993 at Bath and Oxleas Wood, had an especially major impact. For many years there had been arguments about building an extension of the M3 near Winchester, but after a public inquiry at which objections to spoiling a popular piece of landscape had been overruled, plans to build the motorway across Twyford Down were approved. Beginning in 1992 anti-roads protesters encamped on the proposed site and were forcibly ejected by agents of the road constructors. The battle between the motorway construction teams and the encamped 'eco-warriors', as they were depicted, served, as Dudley and Richardson (1998) have said, to enable the anti-roads protesters to use a new arena of political action. Such protests have continued, with figures such as 'Swampy' who sat in trees or tunnelled underground. Swampy's relative popularity is demonstrated by the extent that he became a media celebrity, valued for interviews in tabloid newspapers and appearances on TV shows. The new arena of direct action anti-road protest projected a new image of road building whose environmental destructiveness and socially confrontational overtones was very much at odds with the progressive, modernising image projected by the Department of Transport in support of their road-building programme.

This knowledge about road building was successfully manufactured, disseminated and transformed into a dominant discourse not only because the anti-roads protesters organised their actions in the way that they did, but also because the images and knowledge that they projected had widespread resonance with public opinion, including middle-class opinion, which ministers such as Brian Mawhinney, and a cost-cutting Treasury, were keen to pacify. John Whitelegg, who has been both an academic commentator on the transport debate and a leading anti-road-building campaigner, commented that: 'In addition to the opportunistic possibility of cutting public spending, the Government responded to agitation from their electoral heartlands in places like Wiltshire, Surrey and Sussex showing concern over the environmental effects of roadbuilding.'[1]

It has sometimes been claimed that the actions of outsider protesters, such as those at Twyford Down, or against the Newbury bypass in 1996, are useless rhetorical gestures which fail in their aim of stopping the schemes which prompted the protest. Doubts have been cast on the efficacy of anti-roads direct action protest as a means of altering overall governmental policy on road building. Grant, for example, asks whether cutbacks in road-building plans indicated that the Government was responding to the growing volume of protests: 'Or was it a belated recognition of the fact that building new roads stimulates additional traffic demand...something which points to the potential influence of an informed debate about the limitations of existing policy?' (Grant 1995: 126). However, the argument that road building was economically unsound because it encouraged traffic growth or that new roads did not necessarily create jobs but did create congestion and various social and environmental costs, were not new arguments. Such arguments had been around for many years, at least in the sense that they had been propagated by academics and researchers who questioned the case for road building (Hillman 1976). Indeed, key pieces of research (Purnell 1985; Anderson 1985), upon which SACTRA (discussed earlier) based its findings, had been published well before the *White Paper Roads for Prosperity*. The process by which transport policy was remade in the 1990s did not involve truth being discovered by 'rational' persuasion, but rather new truths were generated and disseminated by emotive actions that resonated with people at the time. The arguments that questioned the economic case for road building are part of a discourse that linked a negative environmental assessment of road building with a less positive appraisal of its economic impacts and suggestions for an alternative transport policy. It was an alternative discourse that, at least in large part, became the dominant discourse of Government policy. This is illustrated by the earlier analysis of the successive *White Papers* on roads and transport policy.

As popular support for the dominant 'road-building' discourse weakened, the notion of politicians' self-interests changed. This led the politicians to choose new policy options, including reducing spending on road building, that more easily fitted in with the altered evaluations of their self-interest.

In making the judgement that direct action roads protesters were instrumental in altering roads policy, we are implicitly recognising that agents can exert power, in the form of the second aspect of power described by Hay: namely, that of the actor having the power to alter the context in which others operate. This is an example of how the Foucauldian notion of de-centred power is limited in scope; however, it is limited in scope, not wrong. The extent of an actor's power to alter the context, in this case the power of the direct action anti-roads protesters, is limited by the degree of resonance that the discourse projected by the anti-roads protesters has with public opinion. In the 1950s people taking direct action against the building of the M1 would most probably have been widely scorned and regarded as latter-day romantics and Luddites. Thus, we can see that the power of an actor to alter dominant discourses is dependent on, is constrained by, structural factors namely the strength and nature of the roadbuilding discourse that the environmentalist is seeking to change. We need both a centred and a de-centred notion of power if we are properly to explain changes in environmental policy, such as the case of road building policy.

Many would argue with this analysis by saying that discursive structures are not the only powerful structural factors. For instance, Marsh (1983) argues that the state tends to defer to the interests of capitalists. This suggests that economic structures are also important. The problem here is that it is impossible to isolate such structures and offer the possibility for definitional closure. Capitalist interests and descriptions of economic structures do not exist outside discourse. The fact that there it great room for a number of interpretations of key elements of allegedly identifiable economic structures (for example, over who is included and who is excluded from membership of the capitalist class) attests to this. However, what we can do is chart how dominant discourses change. Admittedly the study of discourse still leaves room for differing interpretations, but at least there is an undeniable and finite set of, say, *White Paper* texts, upon which to base any analysis. What can be said is that dominant discourses tend to be discourses that are dominant among what are generally agreed to be leading capitalist institutions. Even here, two points need to be made. First, dominant capitalist discourses are not always the same as dominant discourses in government. Second, the perceived interests of capitalists, and others, in relation to environmental policy are part of, and are changed by and through, discourse. Different capitalists will have different perceived interests at different times according to the changing nature of the discourses that are dominant amongst them. In the transport debate, it seems that the perceived (discursively constructed) interests of economically well-heeled middle-class people were reconstructed through discourse.

The utility and limitations of the Foucauldian concept of power/knowledge can also be demonstrated in the case of policy on alternative energy supplies.

Alternative energy and power

Although these days 'alternative energy' is a term that is mainly associated with renewable energy sources such as wind power, solar power and wave power, in the years immediately following World War II the dominant discourse virtually throughout the industrialised world emphasised the 'nuclear alternative'. I want to discuss the politics of the shifting interpretation of the 'alternative' to fossil fuels. In the 1950s the 'nuclear alternative' discourse held that nuclear power would become increasingly important as the alternative energy source to finite fossil fuel resources. For example, according to research by Hall, the UK Government's atomic research establishment based at Harwell said in its 1949 development programme that:

> present difficulties in securing available supplies of coal and a growing public awareness that world reserves of both coal and petroleum are by no means inexhaustible have had the natural effect of focusing hopes upon atomic energy as an economic source of power to supplement or even replace existing sources.
>
> (Hall 1986: 40)

The UK civil nuclear power programme was the first in the world to achieve commercial production of electricity, although the United States quickly overtook the United Kingdom in terms of the construction of nuclear power stations in the 1960s. However, in the 1970s and 1980s the civil nuclear power construction programmes faltered in the industrialised world, with the important exception of France.

The dominance of the civil nuclear power discourse was hardly even challenged until the end of the 1960s, but, during the 1970s and 1980s, an anti-nuclear discourse, promoted in particular by the new wave of radical environmentalists, became more prominent. This discourse consisted of a series of elements that were critical of nuclear technology. These included the alleged links with nuclear weapons production, problems with nuclear waste disposal, excessive costs of the technology, health risks to uranium miners and power plant workers and the association of nuclear power with unlimited growth and material waste (Elliot 1978: 14–15). This discourse was linked with the promotion of renewable energy, as an alternative to both nuclear power and fossil fuels, in the context of an increased emphasis on the efficient use of energy (Flood 1985). Let us look at the case of the UK nuclear power industry first, and then I shall turn to the case of the United States to point out similarities and dissimilarities using the power/knowledge method of analysis.

Although in the United Kingdom orders for nuclear power stations continued to be placed until the end of the 1980s, no nuclear power station has been ordered and built since 1983. How can we use the notions of discourse and power to understand how this happened in the United Kingdom and why it took longer to occur there than in the United States?

By 1995 the post-war practice of government intervention in the economy to support nuclear power had, officially at least, been ended. The UK Government stated, in a 1995 *White Paper*, that:

> The Government cannot identify any reasons why the electricity market should not of its own accord provide an appropriate level of diversity [of sources of electricity supply]. The Government concludes that there is currently no case for intervention in the market in support of additional capacity on diversity grounds.
> (Department of Trade and Industry 1995: 4)

Although the Government recognised the significant contribution made by existing nuclear power stations to targets for reducing emissions of carbon dioxide, it stated that 'The Government concludes, however, that there is at present no evidence to support the view that new nuclear build is needed in the near future on emission abatement grounds' (1995: 3).

Saward has discussed how nuclear power decisions were, until well into the 1970s, taken in the context of a secretive policy community consisting of public and private sector nuclear power interests. All those critical of the dominant 'nuclear alternative' discourse were excluded from the policy community. However, this policy community was gradually eroded, and greater openness was achieved through public inquiries at which there was, among other issues, debate about the risks of radiation raised by groups such as FOE (Saward 1992: 94–6).

In the United Kingdom it was electricity privatisation which heralded the foreshortening of the nuclear power construction programme. The Central Electricity Generating Board (CEGB) was in the public sector and it did not need to justify its nuclear power construction plans through market mechanisms. However, as preparations for privatisation were made, National Power, which had been earmarked to take on the nuclear industry, indicated that it was unable to take responsibility for the nuclear power generating sector. It said it could not justify to its potential private investors taking responsibility for either the costs of decommissioning existing nuclear plant or the investment costs of building new nuclear power stations. Thus, the nuclear sector was taken out of the privatisation process and given to a separately established state-owned company, Nuclear Electric. This was later privatised, having been re-named 'British Energy', with the older nuclear power stations remaining in public hands. Although (at the time of privatisation) the Government did agree a controversial 'fossil fuel levy' on power from fossil fuels to support the nuclear industry, this was earmarked as being for 'decommissioning' of retiring nuclear plant. No money was provided for any new nuclear power stations other than that of Sizewell B, which was already under construction at the time of electricity privatisation. Orders for three proposed large nuclear power stations were cancelled.

Foucauldian notions of power can be invoked in two ways to analyse this turn of events. First, the weakening in the dominance of the 'nuclear alternative' discourse did not dispose the Government, at the time of electricity privatisation, to save the nuclear power construction programme. Nevertheless, the 'nuclear alternative' discourse was still very dominant at the CEGB. The CEGB were able to continue the nuclear power construction programme because of their control over investment strategy (prior to electricity privatisation). Despite the criticisms contained in the anti-nuclear discourse, Government planning inspectors continued to support their plans. Knowledge about the costs of nuclear power and its desirability was controlled by the CEGB and although this knowledge was subject to increasing challenges in the 1980s – for example, at the Sizewell B Public Inquiry in 1983–85 – the CEGB were still able to impose their own set of knowledge-truths.

However, the power–knowledge field of relations changed dramatically with electricity privatisation, for knowledge was now mediated by strict tests of market, rate of return based, commerciality and private sector investor confidence. The knowledge thrown up by the new arrangements was detrimental to nuclear interests because it suggested that electricity from new nuclear power stations was much more expensive and the costs associated with decommissioning old nuclear power stations much higher than previously stated by the CEGB. It was widely believed that the CEGB had been effectively subsidising nuclear power, and it became politically embarrassing for the Government to continue subsidies in the context of a privatised industry. Giving subsidies was also in conflict with the neo-liberal discourse that was fondly embraced by the Government, a neo-liberal discourse which sanctioned the practice of privatising key industries that had been owned by the state.

Contemporaneous with the progressive weakening of the 'nuclear alternative' discourse and the rupture (by privatisation) in the institutional arrangements that had favoured nuclear interests, was the strengthening of a discourse that saw renewable energy as a credible energy alternative to fossil fuels. A 'renewable alternative' discourse, backed by academic, engineering and anti-nuclear interests, suggested that renewable energy could act as a credible non-fossil alternative to fossil fuels. By the time of electricity privatisation the CEGB had sanctioned the building of three experimental wind farms. Renewable energy actually benefited from electricity privatisation since, in order to justify giving a subsidy to nuclear power under the fossil fuel levy arrangements, the Government was obliged to extend this arrangement to renewable energy (Toke 1998). The Government announced a target that effectively meant that renewable energy should be supplying around 3 per cent of electricity needs by the year 2000. By 1995 renewable energy schemes were being subsidised annually to the tune of around £100 million. A Government minister justified the programme on the basis that,

> In addition to market enablement, it will contribute towards the Government's broad aim of working towards 1500 MW of new renewables-based generation capacity in the UK by the year 2000, thereby assisting the Government in meeting its Rio commitment of

returning UK emissions of carbon dioxide to 1990 levels by the same year... [The programme would] ...also contribute to the policy objectives of establishing diverse, secure and sustainable energy supplies and encouraging the development of internationally-competitive renewable energy industries.

(Wardle 1995: 3)

The next Labour Government continued this policy, setting an ambitious target of generating ten per cent of electricity from renewable energy by the year 2010 (DETR 1999: 3). Big players in construction, development and operation of electricity generating plant, including National Power, Taylor Woodrow and McAlpine, which are involved in fossil fuel and/or nuclear electricity generation and construction, became interested in renewable energy and saw their interests as being fulfilled by backing ventures such as National Wind Power.

It is interesting to compare the statements made in support of subsidising renewable energy with those made on nuclear power cited earlier. The promotion of fuel diversity and the reduction of carbon dioxide emissions were adjudged to be reasons for supporting renewable energy, but not for supporting nuclear power. It could thus be said that, in the case of promoting extra electricity generating capacity as an alternative to fossil fuels, a 'renewable alternative' discourse had replaced a 'nuclear alternative' discourse as the dominant discourse. Renewable truths had replaced nuclear truths.

Some would argue that class politics provide at least a major part of the explanation for the fluctuating fortunes of nuclear power. The argument could be that privatisation was planned in order to weaken the miners by opening up the field for combined cycle gas generators to provide an alternative to coal-fired power stations, and that with this being achieved, there was less need to subsidise nuclear power expansion. It certainly appears to be the case that privatisation greatly increased the speed with which gas power stations replaced coal-fired power stations, although it is less certain that the politicians were quite aware that this would be the consequence of privatisation. However, this line of argument still fails to explain why renewable energy was now being subsidised instead. The explanation also fails to explain the fact that the demise of the nuclear power expansion programme has also occurred in most other western states.

The renewable energy alternative discourse achieved dominance at a governmental level, and, flowing from that, a number of non-discursive practices for funding and otherwise supporting renewable energy were established. Yet, opposition from conservative landscape-protection groups prevented many renewable energy schemes from being give planning permission (Toke 1998). The planning sphere has its own set of rationalities which are quite different from the ones affecting policy derived at the DTI, although in the medium term the planning difficulties have influenced policy at the DETR. However, this is a different story, and now I want to summarise my main conclusions concerning the politics of alternative energy.

We can see from this case study that in the United Kingdom control of knowledge of the costs of nuclear power was crucial to its survival as a favoured energy technology. When knowledge was controlled by interests favourable to nuclear power, it continued to be bankrolled by the state, but when this control was lost, and knowledge that nuclear power was expensive was widely disseminated through the market (upon electricity privatisation), the expansion programme was terminated. However, the establishment of the 'nuclear power is expensive' truth does not in itself explain why nuclear power ceased to be a favoured technology, for new renewable energy projects began to be funded just as the nuclear subsidies were being terminated.

In saying this I am not trying to lament the passing of the nuclear power expansion programme, neither do I wish to criticise giving subsidies to renewable energy. My personal views are antagonistic to nuclear power and very much in favour of renewable energy. But I do want to point out that establishing *why* things happen is extremely problematic; discourse analysis avoids answering such questions, restricting itself instead to looking at *how* things happen. No doubt one could argue that renewable energy is now seen as being more in tune with progress than nuclear power, or alternatively perhaps that renewable energy (which comes in relatively small, independent packages) fits in more easily with the post-privatisation institutions, but the causation is indeterminate. Discursive truth, on the other hand, is manifest in the discourses used to justify policies or actions. Such truth is not determined by the merits of intellectual argument or by absolute claims to validity. Rather, these political truths are largely specific to times, places and groups of people; they are justifications for what is useful, and they emerge through discourse in an interaction between ideas and perceived interests, an interaction that results in some discourses being seen as resonant, and others not so resonant.

The UK example of nuclear power may convey an impression that the truth offered by the private sector is always more useful than that of the public sector. However, the experience of the US nuclear power industry ought to provide caution against any over-enthusiastic embrace of the notion that the market can deliver 'objective' truth that is unfettered by the influences of dominant discourses of the time. More than 100 nuclear reactors were put in place in the United States, constituting by far the largest concentration of nuclear power in the world, under the aegis of the private sector. The cost of doing so is still a matter of controversy today. I shall briefly inspect this case.

Truth and knowledge about US nuclear power

The Atomic Energy Acts of 1946 and 1954 established the publicly funded Atomic Energy Commission (AEC), and gave it considerable powers to encourage nuclear development. The AEC used these powers to offer generous subsidies for the design of nuclear power stations and for nuclear fuel costs and, in addition, the Price-Anderson Act, 1957, limited the financial liability of utilities for the consequences of nuclear accidents.

Despite the inducements on offer, the utilities were relatively slow, during the 1950s, in pushing ahead with nuclear development. However, a round of inducements offered by the AEC in 1962 to utilities to order larger nuclear power stations ushered in a rapid period of ordering. Orders for nuclear power stations increased rapidly from the mid-sixties until 1973.

During this period the anti-nuclear movement was relatively so weak that sceptics tended (rhetorically at least) to accept the original premise about the probable inevitability of nuclear power. Rather, they questioned the safety issues. As one sceptic, Senator Thruston B. Morton, put it in 1968:

> No one questions that the development of the ability to create electrical energy through atomic fission is a tremendous accomplishment, or that someday in the distant future we may be forced to depend on it after our other bountiful sources of electrical energy are exhausted or become too scarce and costly to utilize. But we also know that atomic energy is not the panacea of all our energy problems. It was once expected to be, and we are becoming more aware every day of the costs in terms of potential danger to humanity which this proliferating atomic energy program may entail.
>
> (Curtis and Hogan 1980: 43)

This discourse proved to be increasingly resonant with large sections of the US public. The terms of the 1954 Act legislated for public hearings before nuclear power stations could be built, and although such hearings rarely did more than delay the construction of power stations, they were an opportunity for sceptical lobbies to press home their fears about the dangers of radioactive leakages and large-scale nuclear accidents. The Union of Concerned Scientists was one of the most respected lobbies in this regard. The public arguments advanced a discourse that there needed to be stronger safety measures incorporated into nuclear power station designs. The AEC and its successor, the Nuclear Regulatory Commission (NRC), thus accepted the 'knowledge' that progressively higher safety regulations were required. This knowledge even, according to Komanoff (1982), propelled the (pro-nuclear) professional engineering societies to aim for steadily higher safety standards.

Komanoff (1982) has documented how the stricter safety standards and professional engineering practices rapidly escalated nuclear costs. He calculates that in real terms the costs of building nuclear power stations increased by two-and-a-half-fold between 1971 and 1978. Before the Three Mile Island accident, which occurred in 1979, the expansion of the US nuclear power industry was effectively over. Because of the rapid increase in energy prices that occurred in the oil crisis which began in November 1973, projections of increases in future electricity demand were scaled back. The most expensive nuclear power stations on the utility order books, which were invariably nuclear power stations, were the first to be cancelled. No nuclear power station ordered after 1973 has ever been built in the United States. These cancellations tended to occur, in the mid and late 1970s, after evaluations by financial markets rather than by the utilities themselves.

The emerging anti-nuclear discourse promoted the truth that nuclear power stations needed to be built to ever-higher safety standards. However, it also seems that it was never clear that nuclear power stations were competitive on cost grounds compared to the coal-fired and oil-fired power stations available during that period (Komanoff 1982). As early as 1962 the Atomic Energy Commission reported that:

> Power reactors can be reliably and safely operated. However, contrary to earlier optimism, the economic requirements have led to many problems. ... Attempts to optimise the economics by working on the outer fringes of technical experience, together with the difficulties always experienced in a new and rapidly advancing technology, have led to many disappointments and frustrations. ... Many construction projects have experienced delays and financial overruns.
> (Atomic Energy Commission 1962: 4)

As one analyst put it:

> In spite of the optimism and the surge of orders, little solid information was available about [nuclear power station] costs throughout the period of commercialisation from 1965 to 1974. In the late 1960s data on existing plants began to replace sheer estimates, but these data could not be extrapolated to future plants. [That was because the sizes of the plants was increasing so rapidly.] In situations where there are no clear data, everyone must rely on intuitions filtered heavily through worldviews. Technological enthusiasts at the AEC fed optimistic information to technological enthusiasts in the nuclear and electricity industries, who recycled it back. Those responsible for cost-benefit analyses at the utilities had little choice but to accept the assumptions about future developments that the technological enthusiasts fed them.
>
> (Jasper 1990: 48–9)

Thus we can see that nuclear power expansion was predicated on the knowledge of power at competitive prices (at least in the future) from nuclear power. People knew that nuclear power was going to work out. This knowledge was generated by the dominant nuclear power discourse, a discourse that had as its legitimate translators a corpus of nuclear technologists whose word was accepted because they were the servants of a discourse of acknowledged nuclear truths. In terms of Foucault's archaeological method, the nuclear technologists were a key part of the enunciative modality of the dominant nuclear power discourse. In the end, the high commercial costs of nuclear power did emerge through the tests imposed by the market, albeit only after the oil crisis had engendered much closer public scrutiny of energy policy options. However, this case study illustrates how the private sector proceeds on the basis of dominant discourses about what is happening and is likely to happen. Truth does not exist in any latent sense. It is produced.

It is in this context of truth being specific to particular discourses that we must approach the critique of rational choice theory that is the topic of the next chapter.

2
Rational Choice Theory and Environmental Policy

This chapter aims to investigate the extent to which rational choice theory can help to analyse developments in environmental policy. This is crucial for environmental policy, since rational choice theory has grown from obscurity in the 1950s to a theory that promises (or threatens) to dominate political science in the late 1990s. By 1992 almost 40 per cent of articles in the *American Political Science Review* were based on rational choice methods (Green and Shapiro 1994: 3). It is therefore highly relevant to take stock of the possibilities for applying rational choice theory to environmental policy. Environmental policy is a key test for the efficacy of the method. Rational choice theory owes many of its origins to economics, so one key question is whether it can deal with a problem that is outside conventional economics. (Henceforth I shall use the abbreviation RCT to denote rational choice theory.)

A further issue is whether RCT – which, according to proponents such as Dowding (1991: 148),[1] does not seek to analyse the origins of the desires which it takes for granted – can explain and predict outcomes in an issue area where people's preferences have clearly changed (and are still changing). That is, preferences concerning environmental issues have changed in the sense that people want issues dealt with that were once regarded variously as insoluble, as the concern of fringe eccentrics, or as problems that nobody recognised as even existing.

A simple, basic explanation of RCT would be helpful. As Albert Weale puts it (1992: 46), rational choice theory 'is based upon the notion of possessive individualists seeking to achieve their own purposes in competition with one another'. Hugh Ward says, of RCT, that 'arguably the most important tool is game theory' (1995: 77). Game theory involves modelling competitive behaviour wherein individuals seek to maximise their interests. The self-interests of the actors are derived from the attainment of that set of preferences which will maximise the utility (satisfaction) of the actor. RCT is used to predict outcomes, and also to explain past behaviour, although some rational choice analysts would argue that one of the acclaimed advantages of RCT, that of methodological parsimony, is eroded if it is used to explain in a causal sense (Elster 1986: 2).

The usage by modern proponents of RCT has, ironically, to be distinguished from many of the authors who brought it to prominence in the first place. Many of the central ideas (on environmental themes, at least) of 'classic' RCT authors like Olson, Hardin and Downs, seem today, when related to environmental policy at least, to be seriously at fault, and indeed some of the most penetrating critiques have been penned by rational choice adherents themselves. I discuss the classic RCT authors, not because I wish to pretend that these represent what many feel to be the rather more sophisticated variants of RCT, but rather, for two very good reasons: first, because, in order to understand the development of RCT, and some of preoccupations of more contemporary rational choice theorists, we have to understand the classical approach; and second, I want to prepare the ground to answer the question whether, stripped of the bold pronouncements of these classic authors, RCT really delivers sufficient insights into environmental policy and politics to justify it being a (or in many American eyes) *the* pre-eminent basis for analysis of the politics of the environment.

I shall continue this chapter, initially, with an analysis of the problems with classical RCT. I shall then analyse and evaluate the success of the efforts of what I shall call modern, sophisticated rational choice analysts to refashion RCT environmental modelling to deal with these failings. This involves examination of the utility of the notion of 'soft incentives' and the success of Ostrom's use of rational choice institutionalism to revitalise RCT. Finally, I shall move on to a critique of a couple of case studies of rational choice analysis in order to illustrate some of my theoretical points. In the context of this line of argument I shall contrast an alternative 'cultural' approach that embraces the use of discourse theory, which has already been elaborated.

The fall of classic rational choice

The following is a description of what could be described as the 'classical' approach of RCT to the barriers that face efforts to solve environmental problems. First, RCT gives us what I would call the general theory of the failure of collective action.

Garrett Hardin, in his seminal article 'The Tragedy of the Commons' (1968), applied the classic dilemma posed by rational choice theorists; that of the pursuit of rational choice interest against the interests of the collective good. Hardin imagined a herdsman on common-land calculating that the utility of adding each extra animal to his herd would serve his individual interests, even though it was leading to the exhaustion of the common. 'Ruin is the destination toward which all men rush, each pursuing his own best interest in a society that believes in the freedom of the commons. Freedom in a commons brings ruin to all' (Hardin 1968: 1239).

Thus, according to RCT, collective action failure (and also market failure) arises because even though an individual may recognise that there is an environmental problem, they will not contribute to its resolution, because the personal cost of doing so will be larger than the personal benefit that is derived from their own personal contribution to the solution. Even if an effort was made to collect contributions from individual actors to deal with the problem, the individuals may opt not to do so and 'free ride' on the efforts of others to solve the problem (assuming there were any 'others').

The problem which Hardin identified is often formalised into the 'prisoner's dilemma' used in game theory. Game theory attempts to discover the outcome of a situation involving two or more interdependent actors who can take action by making an individual choice either to co-operate to tackle a common problem or to defect (not to co-operate). Each individual action will involve quantifiable payoffs which will be either beneficial or costly to that individual. It is assumed that the actors act rationally to maximise their payoffs or at least to limit their losses.

The prisoner's dilemma scenario is a particular type of situation where the equilibrium position – where the actors make their best response to the set of possible responses by the other player(s) – is one of non-cooperation. Thus, for example, two actors can gain a benefit of 5 units each if they co-operate, but zero each if they do not. However, there is a cost of co-operation, so that if one person 'defects' while the other co-operates, the defector may still pick up net benefits and gain, say, 4 units, while the co-operating 'sucker' ends up with say, minus 2 units. So the 'rational' equilibrium strategy is not to co-operate since, at worst, they at least avoid being 'suckered'. Hence the likelihood of 'collective action failure' under these conditions.

In what sounded like – and was widely interpreted as – an authoritarian message, Hardin called for mutual coercion (albeit mutually agreed) to counter the pillage of the common and ensure that people co-operate to solve environmental problems. Hardin's call for 'mutual coercion, mutually agreed upon by the majority of the people affected' (1968) was echoed by other theorists, such as Ophuls (1973) and Heilbroner (1974).

Hardin's pessimistic-sounding prognoses for environmental problems were paralleled by those of Anthony Downs, author of seminal works on RCT, who also turned his gaze towards what was, in the late 1960s and early 1970s, the emerging field of 'new' environmental politics. He proposed to analyse social problems, taking the example of the environment, by using a model which he called 'the issue attention cycle' (Downs 1976: 29–40). According to this model problems go through four stages. First is the pre-problem stage, where the problem 'has not yet captured much public attention'. The second stage concerns 'alarmed discovery and euphoric enthusiasm', wherein the problem can be countered 'without any fundamental reordering of society itself'. This third stage involves 'realising the cost of significant progress. ...The third stage consists of a growing realisation that the cost of solving the problem is very high indeed. Really doing so would not only take a great deal of money but would also require major sacrifices by large groups in the population.' Fourth is the 'growing decline of intense public interest' as the difficulties and costs of solving the problem are realised; and the fifth phase is the post-problem stage where programmes and institutions may have been set up to deal with the problem and may be delivering some success, but not sufficient to deal decisively with the problem. In effect, this is another way of describing collective action failure. Downs commented (he was writing in the mid-1970s) that the environment problem had already begun moving towards the fourth stage and would eventually move on to the post-problem stage (1976: 37).

As with much classic RCT, Downs elevates a sporadic tendency into an iron law. Often (in practice) interest in the solutions to environmental problems is much thinner than the original alarm, but equally many solutions *do* deal decisively with the problem – the case of lead in petrol is one example that springs immediately to mind. The example of ozone depletion is one that involves active public interest in the solution (banning CFCs), and there is at least a sporting chance that we are also on a path to dealing decisively with the problem. Emissions of sulphur dioxide have been greatly reduced in countries like the United States and the United Kingdom since the time when Downs constructed his issue attention cycle. Downs was clearly wrong about the prognosis for environmental interest generally. Intense concern about the environment re-emerged in the 1980s, and in many ways this concern was more intense than in the previous 'attention cycle' in the 1960s and 1970s. It does seem today that environmental concern has become endemic in western society, and although one could promote a case for a sort of meta-cycles of cycles of environmental concern, then each cycle seems to begin from a higher plateau of concern compared to the beginning of the previous cycle. Downs's analysis is clearly also at fault when he seems to lump all environmental problems into one bag, which makes all environmental issues jump up and down at the same time. Different environmental problems have different cycles, and the cycles – if that is what one should call them – often have patterns that are very different from that presented by Downs. Downs's ideas have, in retrospect, a structuralist ring about them. Hay and Wincott (1998) highlight a general criticism of rational choice methodology for the apparent structuralist assumption that actors inhabiting a given social location have an identical set of preferences.

Besides ozone depletion and lead in petrol – problems that have been the subjects of at least partial collective action success – OECD countries have achieved progress with respect to acid rain and many local and regional problems in the developed world such as some types of car emissions and various types of water pollution (OECD 1993). Wildlife and habitat protection has, in some areas, been achieved through, for example, the ban on commercial whaling and the 50-year moratorium on mining in the Antarctic, which was agreed in 1991. Besides these examples a welter of case studies have emerged, especially since the 1980s, to illustrate how collective action failure with regard to common resource problems is avoided in the case of fishing, irrigation, grazing, forestry and other rights, mostly on the basis of often very long-standing locally run institutions (Ostrom 1990; Birkes and Firkin 1989). I shall return to Ostrom soon.

Another feature of classic rational choice analysis of environmental policy is what I would call the special theory of collective action failure. This was developed in Olson's classic study, *The Logic of Collective Action* (1965). This suggested that it is difficult to mobilise large groups of persons with a common, although individually small, interest compared to small groups consisting of people with concentrated, large interests in a particular outcome. Large groups suffer, according to Olson, from the 'free rider' problem, in that there is a great incentive to let others pay the significant costs to win benefits which may be small in relation to each actor. Such groups could only mobilise people to the extent of joining a pressure group by offering them 'selective incentives' like insurance cover. However, 'privileged' small but powerful interest groups who could look forward to the gains of taking action being more than the costs of political action were relatively easy to mobilise. Olson's ideas rocked the political science world of the time, which had become used to pluralist assumptions about the workings of liberal democracy, since Olson's theory suggested that small, well-organised interests would prevail over the interests of the majority.

These ideas were also a key part of public choice theory, which has usually been associated with right-wing, neo-liberal attitudes. Part of the logic that is said to flow from Olson's study of collective action is that attempts by the state to intervene to promote the public good are likely to be undermined by the tendency of well-organised special interest groups to obtain their own objectives at the expense of the collective good. The only way to avoid this, so the argument runs, is to leave as much as possible to the market and restrict the role of the state to that of ensuring that the market can operate properly and freely.

In fact, Olson said very little about non-economic interest groups and nothing at all about environmental groups. He implied in some places that his statements might not apply to such groups, but nevertheless, the inference from his work was that well-organised, elite producer business interest groups would be much more powerful than environmental interest groups. In fact, Olson's predictions seemed to be contradicted by the empirical evidence of the success of public interest groups in recruiting members, and also (at least on occasions) success in influencing government. Olson's theories could be defended only if it could be shown that public interest group membership was driven by selective incentives. Certainly, one wonders how the environmental movement could have been so successful if the ideas of public choice theorists like Olson were correct. The evidence contradicts Olson's theoretical predictions. A study by Jordan and Maloney (1996) of the membership of Amnesty International and Friends of the Earth in the United Kingdom shows that members were attracted to these organisations principally because they wished to support the campaigning objectives of the organisation, not because they received bulletins or other selective incentives. As Jordan and Maloney put it:

> Our central proposition is that potential members do not engage in the sort of intellectual gymnastics of the economically rational choice approach, costing their contribution against the likely personal return on the investment. Such an exercise may well be

intellectually impossible, because the benefits are difficult to quantify and the sheer effort of the calculation appears to be more than it is worth.

(1996: 669)

So it seems that Olson's predictions about the supremacy of economic motives and the inevitable triumph of the well-organised, privileged (and coerced) groups over the unorganised masses is at least much overstated. Although the conclusions of Hardin, Olson and Downs have been widely attacked, often with great success, and often by more modern rational choice theorists themselves, the aura and influence of the works of these people still remain. In fact, as discussed in Chapter 1, public choice theory as propounded by Olson forms a key part of the neo-liberal discourse that now rules the wealthiest parts of the world. Intellectually, it seems, the neo-liberal discourse, if it were a statue, could be said to have feet of clay at least in the case of environmental policy. Hardin's ideas still have influence in the form of the 'doom and gloom' prognoses that often issue forth. Downs seems also to provide a backdrop to those politicians and others who are cynical about the longevity of popular concern with environmental problems and who believe that if they temporise for long enough then people will lose interest.

It must be said in defence of rational choice theorists that their commentary on environmental policy tended to reflect the feeling of the times. But if RCT can do no better than to reflect the conventional thinking and morality of the hour, its claims to offer predictive insights are hardly justified. So let us examine the potential of modern, new, improved RCT to overcome this problem. Let us analyse some of the arguments of RCT proponents, first by analysing their efforts to criticise classic RCT and rebuild alternative analysis that avoids the classic errors.

RCT and 'soft' incentives

Taking the criticisms of Olson's theory of group membership and his emphasis on economic motives first, some theorists, boxing from what they call a rational choice corner, say that RCT can be amended to take account of motivations, called 'soft incentives', that cannot easily be translated into monetary quantities. Economic self-interest is not the only sort of self-interest. Therefore, the argument runs, so long as these other possibilities for self-interest can be identified they can be incorporated into game theory models of behaviour and predictions of outcomes can still be made.

In the case of the reasons for joining public interest organisations, analysts point to the work of Clarke and Wilson (1961) as a source of a method of categorising non-economic motivations for joining public interest groups. The explanation provided by Rudig et al. (1993: 8) was as follows:

> Material incentives consist of tangible benefits which accrue to group members exclusive of others; solidary incentives arise from the social association with other members; and purposive incentives consist of intangible benefits of membership related to the purpose of the organisation, such as the satisfaction of contributing to the group or serving a good cause.

Rudig et al. used this typology in their analysis of the rise and fall of the Green Party in the 1989–93 period – I say they used this typology in this analysis, but I cannot detect how their use of the typology answers the key question that they set themselves; namely, to uncover 'the factors which could account for the sudden rise and fall of the Green Party'. Now I do regard the study as an effective one in that it did identify clear reasons for the rise and fall in the Green Party's fortunes (rise and fall in salience of environmental issues, change in economic circumstances, political health of the Liberal Democrats) and it did throw up some interesting differences in attitudes between Green Party members and voters. However, none of these lessons are derived from, or even assisted by, the use of the typology of rational interests used by Rudig et al. For example, Rudig et al. found a strong relationship between people leaving the Green Party and blaming the Party for its failures. Yet this, in itself, tells us nothing about the declining salience of environmental issues among the electorate, which was the key factor in the decline of the Green Party if we are to believe Rudig et al's account. Indeed, the only way that the study seemed to use the information on why people are members was to comment on the validity of the typology itself. It was no more, as far as I can see, than an exercise in tautology, a criticism of the use of soft incentives in RCT made by Sabatier (Jordan and Maloney 1996: 677). In fact the information did sustain the explanation of the typology as to why people join parties in general, although of course not why membership expanded and later declined in this instance. The only caveat on the issue of the typology was that quite a few young members seemed to want to join for fashion reasons, which suggests to me that a further soft incentive ought to be added to the typology, that of providing the member with an identity. However, the study did not make this last point.

A similar typology was used by Opp (1986) to analyse people's motivations for going on anti-nuclear demonstrations in Germany, although, again, no attempt was made to use the information to explain the development of the anti-nuclear movement, other than to say that people went on demonstrations because they were opposed, for various reasons, to nuclear power. Asking people questions about why they go on a demonstration is not a technique that is the sole property of RCT, and the skilled mathematical modelling that was conducted seemed to be, as in the Rudig *et al.* study, mainly concerned with a tautological proof of the existence of soft incentives rather than an exercise in analytical enlightenment. Although the study did reveal some useful data on why people went on the demonstrations, it might actually have been an even more crucial piece of research if the study had analysed the dominant (textual) discourses of the anti-nuclear movement to investigate the match between the views of the rank-and-file supporters and those of the leadership.

The concession that economic motives are not the only drivers of rational self-interest (a concession that has to be made because of the empirical evidence) makes things far more difficult for rational choice theorists. Much recent work on RCT is based on often highly complex game modelling to look at how equilibria (conditions when none of the players is worse off by choosing a different strategy) are affected by altering conditions. If the models have to allow for multiple possibilities for pursuing self-interest, thus increasing the numbers of preference functions used, then the process becomes a great deal more complex than it is already. The number of possible equilibria increases exponentially. In short, the ability to make any predictions with any precision whatsoever flies very quickly out of the window. Almost anything becomes possible as a prediction. Explanation of past events becomes even worse, since one does not know which type of self-interest an actor was pursuing at any particular moment in time. It might be argued that a way around this is to limit the RCT analysis to cases where the self-interest can be clearly defined. But does this still leave RCT as a viable mechanism for widespread application, or can it be only used as a useful tool in carefully defined, specific situations?

Rational choice institutionalists, who use rational choice modelling techniques in the context of the study of institutions, do not usually explicitly address this question, but they do attempt to develop RCT modelling techniques that do not suffer the problems encountered by classic rational choice theorists like Hardin.

Ostrom's institutional RCT

The work of Ostrom (1990), mentioned earlier, is widely regarded as an important statement of rational choice institutionalism. Her work represents a great contribution to understanding how the local common has been managed quite successfully for hundreds, sometimes thousands, of years, by ordinary foresters, fishermen and herdsmen. However, I have doubts as to whether it proves that rational choice institutionalism is the key method to analysing these efforts. This is mainly because her own explanation relies, implicitly if not always explicitly, on 'cultural' explanations that are outside the remit of RCT. I shall begin by briefly describing the way she develops RCT into what she describes as an 'institutional' variation.

Ostrom begins her work with a description of five possible games that can be used to model problems surrounding common property resources (CPRs). She reviews them in turn, beginning with Hardin's classic prisoner's dilemma (game 1) and Hardin's suggestion for a 'Leviathan' central agency to impose order on the common (game 2). Although modelling techniques can demonstrate that this game results in the common being saved from ruin, she argues that in practice central agencies have incomplete information, if not about the carrying capacity of the common in question, then certainly about the actions of the herders. A significant degree of uncertainty, over whether penalties for breaking rules preventing bad use of the common will be enforced, is incorporated in 'game 3'. Because free riders feel they have more chance of evading penalty, this game results in another prisoner's dilemma situation and collective action failure. This demonstrates that Hardin's 'Leviathan' solution will not work to save his imaginary common.

Ostrom discussed another 'centrally imposed' solution – that of privatisation – in 'game 4'. Although it harnesses rational choice motivations in that owners of resources are determined to take action to prevent their becoming degraded, and Ostrom acknowledges that this technique might be useful in some circumstances, its claims to be the only solution are often undermined by two general factors. First is the extreme difficulty of dividing up some resources. Fisheries are cases in point. One could also make the same point about air pollution problems in general. The second difficulty is that often grazing the common requires co-operative behaviour to exchange resources which are owned in an uneven fashion by the different 'herdsmen'. Ensuring co-operation and substituting for it by insurance can be very expensive under private ownership.[2]

It is what Ostrom sees as 'game 5' that contains the kernel of the recipe for collective action success. When a CPR institution is locally managed, free riding is much more likely to be detected because the locals can devise much more effective surveillance methods than the central authority. In fact much of what Ostrom says is music to the average member of the modern decentralist green tendency. She says that in practice, successful efforts to manage CPRs in a sustainable manner are generally organised by local people themselves according to rules which are locally determined. She identifies seven important design principles: (1) a definition of who has rights in the CPR; (2) rules governing what, when and how things may be taken from the resource; (3) participation by appropriators in deciding rule changes; (4) monitors are either appropriators themselves or people who are accountable to them; (5) the application of sanctions to people who break the rules by the appropriators or their officials; (6) conflict resolution mechanisms; (7) recognition by external authorities of the rights of appropriators to regulate their CPR institution (1990: 90). Effective monitoring of behaviour and certainty of penalty for defecting from the rules leads to a game theory analysis of how this produces an outcome where people choose the co-operative strategy. Under these circumstances the co-operative strategy is the one which is likely to maximise payoffs to each actor. It seems, therefore, at this reading, that RCT provides an effective explanation of collective action success in the case of CPRs if it is combined with the right institutional theory. Institutions, according to rational choice institutionalists, constrain the actions of individuals by altering their incentive structures. The costs and benefits of an action are altered by institutional rules, thus potentially altering the choice of action – in particular choices about whether to 'free ride' on resources or to act to support the collective good. Or, as Hall and Taylor put it:

> [Rational choice institutionalists hold] that an actor's behaviour is likely to be driven...by a strategic calculus...[which] will be deeply affected by the actor's expectations about how others are likely to behave as well. Institutions structure such interactions, by affecting the range and sequence of alternatives on the choice-agenda or by providing information and enforcement mechanisms that reduce

uncertainty about the corresponding behaviour of others and allow 'gains from exchange', thereby leading actors toward particular calculations and potentially better social outcomes.

(1996: 945)

It might seem, therefore, from this account so far that institutional RCT has succeeded in modernising RCT by linking it to study of institutions. But there are persuasive arguments to suggest that this can only be done by borrowing from what Hall and Taylor call a 'cultural' approach. This is in contradistinction to the 'calculus' approach used in rational choice modelling, which assumes that individuals are instrumental utility maximisers. The cultural approach 'stresses the degree to which behaviour is not fully strategic but bounded by an individual's worldview... it emphasises the degree to which individuals turn to established routines or familiar patterns of behaviour to attain their purposes' (Hall and Taylor 1996: 939). According to Hall and Taylor, a cultural approach is employed by sociological institutionalists and also by historical institutionalists, the latter pursuing an eclectic mixture of calculus and cultural techniques together with an understanding of the implications of path-dependent development. Hay and Wincott (1998) develop what they see as a distinctive model of historical institutionalism. While they stress the importance of path dependence, their model is not merely a pastiche of 'calculus' and 'cultural' methods but one which involves a dynamic interaction between interests and institutions. Agents operate in an environment of uncertainty about outcomes, and a likelihood that unintended consequences will follow their actions. Moreover, their notion of self-interest is apt to be reinterpreted by changing structures. This seems a useful model to adopt if we assume that the structures in question are discursive structures.

This historical institutionalist approach has been used, for example, to analyse the creation of an energy-efficiency policy network in the UK buildings policy sector (Toke 2000). A crucial point made in this article, in contradistinction to rational choice analysis of policy network creation (Blom-Hansen 1997), is the role of the cognitive structures. Cognitive structures are the views of goals and reality associated with a particular problem. They arise from discourse and will, in terms of policy communities concerned with government decisions, consist of dominant governmental discourses which will underpin the policy network. The cognitive structures influence the 'decision rules'; namely, the rules for interaction between network participants. Hence, for example, if there is a dominant world view, a dominant discourse, accepted by network participants, then the network is likely to be co-operative and coherent in nature and tend towards being a policy community rather than an issue network, as defined by Marsh and Rhodes (1992) and Daugbjerg (1998b). This is further evidence that rational choice institutionalism cannot be used as the sole basis for examining institutional formation and persistence, a point that can be made through careful analysis of Ostrom's own model.

Belief systems, norms and self-interest

A careful reading of Ostrom's own arguments tends to undermine claims that she or others may make to establish rational choice instituionalism as *necessarily* providing the principal means for explaining collective action success. Implicitly, and often explicitly, she describes the importance of norms and belief systems to the formation and perseverance of the institutions that manage the CPRs under discussion.

Ostrom identifies four internal variables governing individual choice: 'expected benefits, expected costs, internal norms and discount rates' (Ostrom 1990: 37). It should be noted that discount rates, costs and benefits must, according to her argument, all be influenced by the norms. Norms are, in Ostrom's model, both internalised and influenced by external normative structures. Ostrom observes how discount rates are affected not only by levels of physical and economic security, but also 'by the general norms shared by the individuals living in a particular society or community regarding the relative importance of the future compared with the present' (1990: 35). In general, she says:

> Norms of behaviour affect the way alternatives are perceived and weighed. For many routine decisions, actions that are considered wrong among a set of individuals interacting together over time will not even be included in the set of strategies contemplated by the individual. If the individual's attention is drawn to the possibility of taking such an action by the availability of a very large payoff for doing so, the action may be included in the set of alternatives to be considered, but with a high cost attached.
>
> (Ostrom 1990: 35)

The logic of this argument casts doubt on the tendency of rational choice analysts to separate out norms and self-interest. Elster, who recognises an influential but subsidiary role for norms in comparison to rationality, comments, 'we do not know what determines when norms remain strong and stable and when they yield to the pressure of self-interest' (1986: 24). This distinction is also maintained by Ward, who reports the view that 'people are more likely to conform to norms when conforming has low costs' (1995: 85).

The model of individual choice used by Ostrom implies that all the relevant decisions are shaped by norms, something with which analysts following the 'cultural' approach would agree. Many contemporary 'culturalists' would argue that there exist a multiplicity of norms and world views acting on and influencing the individual's sense of self-interest, often acting to pull the individual in conflicting directions. According to this view, which I share, all our choices are influenced by cultural factors, whether it be buying a car, deciding if and when to marry, deciding whether to adopt what are discursively constructed as 'conventional' or 'alternative' lifestyles or, more mundanely, deciding whether to vote in an election or spend the time gardening instead. Cultural influences are not things we can ignore or adopt at will; they profoundly influence our choices in one direction or another. One of Foucault's most trenchant points was to say that it was impossible to conceive of a person as a subject that is free from social conditioning, 'One has to dispense with the constituent subject' (1980: 117, more fully quoted in Chapter 1). To imagine that there is a rational self-interest that is entirely independent of norms and other cultural influences is to forget that our attitudes, including the nature of our self-interest, have been shaped by our culture ever since we sat on our mother's knee for the first time.

Crucially, Ostrom points to the inadequacy of RCT in explaining why appropriators overcome disagreements about what rules there should be in order to set up the institution in the first place. She quotes Bates, with whose arguments she agrees, as finding this conundrum 'deeply puzzling'. Bates does establish the possibility for doing this under game theory, but only after 'establishing trust and a sense of community'. After the institution is set up it operates not merely through altering the incentive structure of the appropriators but also because shared norms will reduce the amount of opportunistic behaviour from appropriators (farmers, fishermen and so on) and 'thus reduce the cost of monitoring and sanctioning activities' (1990: 36). Foucault would have much to say about the use of surveillance by local people to achieve knowledge, and therefore power, to protect their common resources. Dryzek has much to say about how Ostrom's institutions are developed through a 'learning process':

> Many of her cases have persisted for hundreds of years, and current institutional arrangements are the products of lengthy periods of learning about what practices produce good results for the commons and the commoners and what do not. This learning need not be done in any particularly self-conscious way, and it may take the form of accretion and modification of traditions, or cultural practices relying on something like natural selection.
>
> (Dryzek 1996: 35)

In other words, it is impossible to explain the setting up and operation of institutions to manage CPRs without the use of what Hall and Taylor would call 'cultural' explanations that rely on understandings that are shared by the community involved. But the contribution of 'cultural' theory does not end here. Ostrom makes clear that judgements concerning the need for new arrangements do not arise as facts from nowhere; they have to be interpreted: 'Benefits and costs have to be discovered and weighed by individuals using human judgement in highly uncertain and complex situations that are made even more complex to the extent that others behave strategically' (1990: 210). The process of weighing up costs and benefits, being an interpretative process, cannot satisfactorily be modelled by game theory since the costs or benefits associated with a particular strategy or outcome are interpreted through discourse. Analysis of discourses is the only form of knowledge that is clearly available about these historical processes, assuming of course that it has been recorded in some form.

Ostrom describes the single most important characteristic of appropriators achieving success in collective action to be that 'Most appropriators share a common judgement that they will be harmed if they do not adopt an alternative rule' (1990: 211). In other words, it is crucial that there is common world view about the existence of a problem and the need to take collective action to deal with that problem, or (to use the vocabulary explained earlier in this book) there has to be a dominant discourse concerning the need for the institution.

It should be noted that in her study Ostrom is talking mainly about problems whose existence have been recognised for many hundreds of years. The importance of the common world view concerning the existence of a serious problem and the need to take radical steps to deal with it is thus partially obscured by its distant inception as a dominant discourse and the fact that today this common world view is taken for granted when once, long ago, it was a matter of uncertainty. Yet this is the most fundamental aspect of the institution; it is its whole *raison d'être*, and it is also the subject about which RCT is least able to contribute to explanation, precisely because RCT disavows interest in the origin of actors' preferences. This shortcoming becomes more visible when we analyse issues and problems that are relatively new in nature, which puts RCT in a particularly poor position to analyse policy relating to the series of new pollution and conservation issues that have emerged since the 1960s. The most important point about collective action success in dealing with such issues is that people agree that a problem exists, that action needs to be taken to solve this problem, and that a world view exists about feasible technical solutions to this problem. Only then can institutions be devised that can tackle the problem effectively. It is possible to conceive of different sets of institutional arrangements that can deal with an environmental problem: see for example, the successful use of both regulatory and emissions trading regimes to deal with acid rain. But without a secure agreement that an environmental problem exists and has to be addressed, then the institutions necessary to deal with them cannot and will not emerge.

Although work such as that carried out by Ostrom does indicate that RCT has an important role in studying how institutions shape preferences, the argument above clearly indicates that there is a lot more to institutional study than the calculus approach. Institutions formed to alleviate environmental problems can be fully understood only if there is a study through discourse of the world view, the cognitive basis, that underpins the institution. As I argue elsewhere (Toke, 2000) in the context of policy network theory, the more coherent the cognitive structure of the institution is, then the more the institution is likely to command greater co-operation among its participants. Thus institutional analysis cannot be claimed as territory that belongs solely or even mainly to rational choice theorists. Moreover, the previous discussion on the culturally influenced nature of self-interest means that rational choice analysis based on game modelling can produce useful predictions only if the nature of self-interest is very carefully specified. In addition, any predictions made through game modelling of these carefully specified sets of self-interests can be reliable only if we assume that the world views that influence the actors remain constant. Of course, in a typical political situation it is often very difficult to specify the self-interests of the actors, and it is unlikely that belief systems are going to remain constant over more than very short periods of time.

These shortcomings of RCT can be demonstrated by looking at efforts to use RCT to explain how outcomes in environmental policy have been achieved and by examining efforts by RCT to make predictions about the future in cases involving international and global environmental problems. The problems considered here are different from the ones considered by Ostrom, not only because they are international in nature, but also because world views on them have changed, are changing and thus seem likely to change in the future.

RCT and new international environmental problems

When I say that problems like acid rain, global warming and ozone depletion are relatively new, I do not mean necessarily that their physical manifestations are new, but that the widespread recognition of them as problems is new. It is crucial to understand this when one analyses Sandler's efforts (through RCT) (1992) to use game theory to analyse problems like ozone depletion and global warming. These problems are characterised, respectively, as games of different types. Ozone depletion is characterised as a game of 'privilege' where the gains to each actor (in this case nation) outweigh the costs of taking action. On the other hand, global warming is characterised as a classic prisoner's dilemma where collective action failure is taking place since people are deterred from taking action because the costs of such action to the individual nation clearly outweigh the gains.

In fact, such pronouncements suggest that in common with earlier, more pessimistic, analyses by classic rational choice analysts, RCT has a tendency to reflect existing world views or dominant discourses concerning the nature of the problems. This is apparent from statements like 'Even in the heaviest user [of CFCs] nations there are grounds to believe that the net gains from phasing out CFC emissions is positive' (1992: 19). Note the phrase 'grounds to believe'. Progress has been made because of changes in the dominant discourse concerning ozone depletion. The first shift was away from ozone depletion being an unknown issue. Then, in the 1970s, in some countries at least, it was widely regarded as a problem largely in the minds of a few scientists and campaigners that would cause great cost to certain industries to 'solve'. Then, after revelations about the appearance of a 'hole' in the ozone layer, this changed, and 'cheap' solutions were found to the problem. In other words, the problem became soluble because belief systems had changed, not because it was constructed as a game of privilege in the first place.

Similarly, the construction of the global warming problem as a prisoner's dilemma merely reflects the dominant belief system that exists at the time of writing. The 'prediction' made by Sandler is that international agreement on global warming 'might be more achievable prior to the identification of the gainers. Once this uncertainty is resolved, strong vested interests materialise to oppose curbing emissions.' This seems to me to be a little bit of an odd statement, implying that environmentalists' best hopes rest on keeping people in ignorance about the consequences of global warming. It does seem to be counter-intuitive in an era when greens seem to make progress in the wake of scientific revelations rather than ignorance. The discoveries are often presented as apocalyptic stories of the doom that awaits us if we do not take urgent action.

However, in analytical terms, the central problem with Sandler's prediction is that it is dependent on the maintenance of the current dominant world view, which holds, essentially, that solving the global warming problem (which may not really be as bad as the doomsters say) is going to be a costly business. In fact many environmentalists, including the well-known soft energy exponent Amory Lovins, dispute this world view. He believes that the key problem is lack of knowledge about how energy efficiency can be implemented to save money. He says that carbon dioxide emissions can be greatly reduced without the need for high energy taxes. In a publication produced just before the 1997 Kyoto Conference on cutting emissions of greenhouse gases he attacked Robert Samuelson's assertion that politicians faced a dilemma in that it would be political suicide to take action. He continued:

> The dilemma arises because almost everyone presumes that protecting the climate will be costly. ...Samuelson, like so many business people, believes climate protection is costly because the best known economic computer models say it is. Few people realise, however, that those models find carbon abatement to be costly because that is what they assume. This assumption masquerading as a fact has been so widely repeated...that it's often deemed infallible. It's not. Not only do other economic models derive the opposite answer from different assumptions, but on an enormous body of overlooked empiricism...shows that the technological breakthrough Samuelson seeks have already happened. The Earth's climate can be protected not at a cost but at a profit.
>
> (Lovins and Lovins 1998: 1)

Of course, Lovins' analysis may prove wrong, but his ideas are still plausible. The problem with Sandler's analysis is that it implicitly assumes they are wrong. The key point to understand here about Sandler's RCT-based predictions is that they are valid in situations only where the world views do not significantly change, and where self-interests are also constant and easy to describe. In environmental policy, solutions are found because world views emerge that reconstruct the problem and that act to reconstitute the perceived interests of the actors. Unfortunately for RCT, it cannot analyse or incorporate such shifting conceptions of dominant world views and senses of self-interest without reducing its predictions to a conclusion that anything, or almost anything, can happen.

The 'facts' upon which RCT places so much importance in order to assess payoffs of different strategies turn out to be very slippery in the case of global warming. I have discussed the tremendously differing assessments of the costs of dealing with the problem. There is also tremendous disagreement about the damage likely to be caused by the problem. Nordhaus's estimates of the damage costs per tonne of carbon emitted come to $7.3 (Nordhaus 1994). A study commissioned by the EU (Eyre *et al.* 1997) came to very different figures, citing a range of $50 to $170 of damage costs per tonne of carbon emitted. Both Nordhaus and Eyre base their estimates on UN-sponsored Intergovernmental Panel on Climate Change (IPCC) estimates of the likely temperature rises. Demeritt and Rothman (1999) suggest that IPCC estimates of costs and risks of future climate change have been subject to political manipulation.

Although environmental economists such as Pearce *et al.* (1989, 1993) do not engage in game theory, their attempts to monetarise environmental problems do represent a means by which rational choice theorists can ground their notions of payoffs for taking different actions. But as the above discussion reveals, this is very problematic, and the results of attempts by environmental economists to value environmental damage and the costs of repairing that damage seem to be very dependent on the methodology adopted, not to mention prevailing (but changeable) world views of the people they are talking to when they ask them how much they value an environmental asset. Asking people how much they would pay to preserve an environmental asset is part of a technique known as 'contingent valuation'. Jacobs, in a critique of the neo-classical economic approach to environmental policy, attacks contingent valuation techniques (which are used a lot by Pearce *et al.*) on the grounds that it 'asks the wrong question. Asking the personal question "How much are you willing to pay?" encourages people into a self-interested stance. This is the appropriate question in a market for a private good' (1997: 217). Policy is thus decided on the basis of what individuals as private consumers want rather than what citizens who are members of a collective seeking to defend a collective good want. Criticisms such as this make it easy to understand how rational choice analysis can lend itself to anti-collectivist public choice interpretations, since it focuses on the preferences of the private individual operating in a market situation rather than the preferences of a citizen as part of a social collective.

Of course, not all rational choice analysts follow the public choice variant. Ward (1996) is one who does not, but that does not necessarily commend his rational choice analysis of climate change negotiations, although he does come to conclusions that sound rather less banal than some of Sandler's predictions. He bases his analysis on similar assumptions to that of Sandler concerning the high cost of tackling global warming. At least Ward recognises that some argue that there are cost-effective energy efficiency savings. Yet his guiding judgement on costs still seems to be: 'However, major reductions in emissions are predicted to be costly' (1996: 851). He also comes to fairly conservative conclusions – that it will be difficult to agree radical cuts in emissions among a large group of nations – confirming my assertion that rational choice analysts tend to confirm prevailing orthodoxy in the guise of giving apparently objective predictions.

Ward tries to argue, based on a considerable amount of game theory mathematical modelling, that the idea of trying to commit a few countries to co-operate in agreeing to relatively high levels of reductions in carbon dioxide is unlikely to draw in large numbers of other countries. He comments:

> The assumption is that once a high level of co-operation is firmly institutionalised, other countries will be pulled in. However there is no good reason from a game-theoretic perspective to suppose that the coalition would eventually grow to include all the major players, as is typically assumed. Given that there are likely to be equilibria where some nations co-operate and others do not do so at all, a point will be reached where it does not pay additional nations to join the group of ambitious pollution reducers.
>
> (Ward 1996: 865)

It is indeed, a gloomy prognosis, given that the IPCC says that radical cuts in global emissions will be needed to stabilise the climate, and the only hope Ward seems to offer is that trade sanctions could do the trick, although this would require changes in world trade rules. I do not think that trade sanctions are likely over this issue, but then two central RCT assumptions made by Ward are deeply flawed. The first flaw is one I have stressed earlier; namely, that Ward assumes that there will be no major change in belief systems concerning the costs of global warming abatement. He tries to cover himself against this charge by commenting in the conclusion that 'Game theory cannot constitute a free-standing explanation because it takes states' preferences, beliefs and strategic opportunities as givens'. Given that environmental problems only ever become solved either because technical changes make the problem obsolete or because people become much more concerned than they were about the problem and demand urgent action (or often a mixture of the two), then Ward's analysis becomes an exercise with severe limitations since he is looking at conditions which are extremely restrictive.

Ward's conclusions appear to criticise the actions of environmentalists and scientists who are working for countries to adopt radical greenhouse reduction targets, and part of the point of trying to commit countries to do that is in order to persuade other countries to follow suit later. In the case of the development of international accords on reducing acid rain emissions, the strategy of beginning with a small group of countries agreeing significant level of emission reductions in the early 1980s did lead to deeper agreements on cuts being made later on among a large group of nations (McCormick 1997). Getting some countries to make radical commitment and take radical action not only has a potential impact through changing belief structures, but it also has the potential for knock-on effects of pushing technology into more sustainable ways. Hajer analysed how British environmental policy was discursively reconstructed in the 1980s following pressure from Scandinavia to allow modest measures to reduce acid emissions (1997: 106–11). In the case of acid rain, the relative cheapness of acid abatement technology was demonstrated by the early leaders in efforts to cut emissions; and in the case of global warming, even the relatively weak commitments to cut emissions made so far have encouraged greater investment in renewable energy technologies which are making them noticeably cheaper. RCT analysis omits inclusion of such phenomena in its modelling, and consequently its predictions are unreliable.

There is a second key flaw in Ward's analysis (typically reflective of RCT): namely, that nations assess their self-interest on issues such as global warming using a calculus type of approach. The fact that it is difficult to specify self-interest in many situations is a problem with RCT that we discussed earlier. Ward even discusses the possibility that countries might retaliate against non-cooperators and deliberately allow their emissions to increase (1996: 856). But, in the analytical strategy used in this book, in the case of a state that was concerned to combat climate change, national interest is identified with the pursuit of a dominant discourse on combating climate change. This is a cultural phenomenon. The notion of 'defecting' in a game-theoretic sense is foreign to this discourse since the discourse consists of a series of elements, which includes the idea of the need to co-operate with other nations to reach targets and the need to work consistently to gear domestic policies to reduce emissions. Greenpeace (or even the Government) is not suddenly going to tell citizens to buy inefficient refrigerators or large, 'gas-guzzling' vehicles as a means to retaliate against some other nation which will not agree to aim for bigger emission reduction targets. Such an action would, in any case, be intuitively ineffective since the 'defecting' country is defecting precisely because it is relatively unimpressed by the global warming discourse. Hence it would not be impressed by action to punish it by action (such as allowing emissions to rise) which it thought would not make much difference anyway.

Paterson makes criticisms of rational choice assumptions that are similar to some made by me when he says, in criticising realist accounts of global warming:

> When examining states' actions, an interpretation which suggests that they are rational actors in the sense outlined by RCT is less plausible than an interpretation which suggests that they are role players and reflexive about their goals. ... [They are not] rational unitary states instrumentally pursuing their predefined national interest.
>
> (Paterson 1996: 178–9)

The verdict of this study of the efficacy of rational choice as a tool for analysing environmental policy must be that the disavowal, by rational choice theorists, of the intrinsic and congenital, rather than occasional, importance of cultural influences puts it at a disadvantage in analysing developments in environmental policy. This is not to say that institutional rational choice theorists cannot bring important insights, but if Ostrom's work on CPRs is typical, these gains will be produced only by combining the 'calculus' approach with other techniques borrowed from other schools of thought, especially those associated with what has been referred to as historical institutionalism. Game theory may provide insights in situations where world views remain constant and the self-interests of the actors can be very clearly defined in a unidirectional way, but if the examples discussed here are typical, its usefulness is limited when it comes to analysing environmental policy.

My criticism of RCT does not mean that the agency side of analysis is ignored, a point which I hope emerges from Chapter 1. Indeed, I now, in the next chapter, want to investigate the roles of the two key agencies of what I will describe as the modern environmental discourse and their impact on scientific discussion of environmental policy. These key agents are scientists and environmental groups.

3
Science, Politics and Environmentalists[1]

In this chapter I want to look into the way that the development of the modern environmental discourse has influenced the relationship between political discussion and science. A key part of my efforts to do this will be to look at the relationship, to one another, of agents of the environmental discourse, particularly scientists involved in ecology and environmental pressure groups. This helps us understand how, in general terms, current debates about environmental issues are organised and influenced, and how things have changed since the Second World War.

I shall begin with a description of some essential aspects of the modern environmental discourse and I shall then discuss the emergence of this discourse in the post-Second World War world. I shall then relate this to discussions about politics being scientised. I shall then move on to a discussion of the relationship of the agents of the environmental discourse, scientists and environmental groups, which I shall analyse through a critique of the theory of epistemic communities.

The dominance of what I would call the modern environmental discourse, saying that human activities threaten the future of at least humankind on the planet, was signalled by the Brundtland Report (WCED 1987), a report that was soon to be endorsed by the governments of the major industrial nations and given further substance by the United Nations Conference on Environment and Development held in Rio de Janeiro in 1992.

The statement of a minister in the Government of Zimbabwe, reproduced in the Brundtland Report, gives the main elements of the newly dominant discourse:

> The remarkable achievements of the industrial revolution are now beginning seriously to be questioned principally because the

environment was not being considered at the time. It was felt that the sky was so vast and clear nothing could ever change its colour, our rivers so big and their waters so plentiful that no amount of human activity could ever change their quality, and there were trees and natural forests so plentiful that we will never finish them. After all, they grow again. Today we should know better.

(WCED 1987: 34)

The notion that human activities were threatening the ability of the world's ecosystems to support humanity in the future led on to the concept of sustainability, the notion that we must sanction only those activities that could be continued indefinitely without threatening the future of, at least, humanity. The extent of the conversion was signalled when Mrs Thatcher endorsed sustainability in a speech in 1988: 'We Conservatives... are not merely friends of the Earth – we are its guardians and trustees for generations to come. ... No generation has a freehold on this Earth. All we have is a life tenancy – with a full repairing lease' (McCormick 1991: 60).

The modern environmental discourse thus contains at least three essential elements: criticism of industrialism's failure, hitherto, to take the environment seriously; the notion that human activities are threatening human futures; and the concept of sustainability. The political mainstream would interpret sustainability to be acceptable in the form of sustainable development, the principle of endorsing the attainment of economic development which is compatible with sustainability. The attachment to economic development is not wholeheartedly endorsed by the entirety of the green movement, although, as Hajer argues, the idea of ecological modernisation – that the goals of environmental protection and economic development are not only compatible but also mutually reinforcing – is accepted by mainstream environmental campaigners such as Friends of the Earth.

Beyond the basic elements of the modern environmental discourse there is considerable divergence. Rachel Carson's ideas, which were quite firm in denouncing the human project of controlling nature, as opposed merely to countering the threatening effects of this project, have still not achieved discursive dominance among the majority of the world's states. Carson expressed this point of view evocatively (in the context of her critique of pesticide use) in the very last paragraph of *Silent Spring*:

> The control of nature is a phrase conceived in arrogance, born of the Neanderthal age of biology and philosophy, when it was

supposed that nature exists for the convenience of man. The concepts and practices of applied entomology for the most part date from that Stone Age of science. It is our alarming misfortune that so primitive a science has armed itself with the most modern and terrible of weapons, and that in turning them against the insects it has also turned them against the earth.

(1962: 257)

As Dryzek's treatment of environmental discourses demonstrates (1997), there are a number of different variations of modern environmental discourses. There is and has been general agreement on the contribution of the importance of technology to the ecological crisis, but less agreement on the contribution of population and economic growth. Ehrlich developed the I=PAT formula which suggested that environmental impact was a function of population, affluence and technology (Ehrlich and Ehrlich 1990), but others, such as Commoner (1972) disputed the importance of the population and economic growth factors, and argued that the prime task was to improve our choice and deployment of technology.

Although I do not want to repeat efforts to summarise these conflicting interpretations, I do want to move on to discuss the emergence of the modern environmental discourse, because this will put us in a better position to consider the relationship between science and environmental policy.

In the shadow of the bomb

Although there has been a recognisable environmental movement since the mid to late nineteenth century, this was concerned much more with nature preservation and aesthetic criticisms of industrialisation than the general threat of pollution. Paelhke (1989: 21) said of the modern environmental movement, which emerged in the 1960s, that 'The new concern about pollution was immediate, basic, urban; it even crossed class boundaries. Not everyone has the time and money necessary to appreciate wilderness, nature at a distance. But everyone eats, drinks and breathes.'

It is possible to chart, in the period leading up to World War II, an increase in expectation that problems such as urban smogs should be combated (Ashby and Anderson 1981). However, the detonation of nuclear weapons at Hiroshima and Nagasaki, and the subsequent fears about radioactive fallout from atmospheric nuclear testing, constituted perhaps the most important watershed for the modern environmental discourse of which Carson was a leading agent. According to Weart (1988: 325), Carson was inspired to begin writing *Silent Spring* in the mid-1950s:

> She wrote privately that in earlier years, despite scientific evidence about harmful chemicals, she had clung to the faith that 'much of nature was forever beyond the tampering reach of man...the clouds and the wind and the rain were God's'. It was radioactive fallout, she said, that had killed this faith. *Silent Spring* opened with a fable of a town dying from a white chemical powder that Carson likened to fallout.

There is evidence that the emergence of the new, post-atom-bomb discourse regarding human's ability to destroy the natural basis upon which humans and other species alerted, for the first time, a number of people to the possibility that global warming could pose a major ecological threat. Arrhenius had suggested a link between carbon dioxide levels and the Earth's climate in the previous century. There were temperature rises in the 1930s, and Callendar (1938), building on the work of Arrhenius, said that increases in the levels of atmospheric carbon dioxide could lead to global temperature rises. Callendar did not actually suggest that this was an environmental problem[2] as he said that the temperature increases would be benign in nature. This was despite the experience of the drought in the Midwest in the 1930s that had caused so much social misery.

However, after World War II the prospect of global warming represented, to the small but growing group of scientists interested in the issue, a much more sinister development. These fears were to lead to the setting up of a facility, in 1957, at Mauna Loa in Hawaii, to ensure permanent monitoring of atmospheric carbon dioxide levels in the atmosphere. Fleming (1998: 118–19) talks of the emergence of fears, in the early post-World War years, that the testing of nuclear weapons could affect the climate. The nuclear scientist, Edward Teller, for example, became worried during the Second World War that the nitrogen in the atmosphere could be ignited by a fusion bomb (Rhodes 1995: 254). Today this linkage seems incorrect, but the fears came out of the same mind-set as that which was held by those who were worried about the consequences on the climate of increasing carbon dioxide emissions. The point was that the advent of the tremendous and terrible power of nuclear weapons made people realise that human machines had the ability to devour their makers. This message was carried in the new environmental discourse. The immediate significance (for environmental policy) of the post-war debate about the dangers of nuclear weapons and nuclear fallout from atmospheric testing of nuclear weapons (which was banned in 1963) has subsided in many people's minds. However, the significance of this debate lives on in the way it crystallised the formation of the modern environmental discourse. It is this discourse that formed the backdrop to the way increasing numbers of scientists and activists became sceptical of a range of industrial practices.

From the scientisation of politics to the politicisation of science

It has been said that Habermas and other critical theorists have not been influential, in a direct sense, on the thinking of the green movement (Eckersely 1992: 99), yet the common influences on environmentalist and Habermasian attitudes to the malevolent control of knowledge by industrial interests are clear. Habermas' earlier writings are particularly concerned about western society becoming 'scientised'. This was said to involve vested interests using a cloak of scientific legitimacy to avoid public debate on the dangerous nature of the application of their technologies (Habermas 1971). Yet such worries sprang from the same concerns about the danger of nuclear weapons and nuclear fallout that activated many intellectuals to speak about human ability to damage nature. Habermas challenged the notion of scientific positivism, which said that the world could be fully described through a scientific search for objective absolute truths. He alleged that establishment experts were using their dubious positivist claims to possession of absolute truths to further certain powerful military and industrial interests.

But was Habermas merely analysing a growing trend towards 'scientisation' of policy making, or was he, through his critique, at least as significant in providing intellectual legitimacy for a growing trend in the West for people to be sceptical about the automatic benefits of scientific progress? Weart, for example (1988: 323–4) reports opinion surveys in the United States as saying that even in 1966 only half the respondents said they had 'a great deal' of confidence in science and that this proportion had fallen to a third in 1973. Habermas did point to some signs that rebellion, even among scientists, was beginning to stir:

> Since [the invention of the atomic bomb] there have been discussions in which leading scientists have argued about the political ramifications of their research practice, such as the damages that radioactive fallout have caused to the present health of the population and to the genetic substance of the human species. But the examples have been few and far between. ...The question is...whether a productive body of knowledge is merely transmitted to men engaged in technical manipulation for purposes of control or is simultaneously appropriated as the linguistic possession of communicating individuals. A scientised society could constitute itself as a rational one only to the extent that science and technology are mediated with the conduct of life through the life of its citizens.
>
> (1971: 79–80)

But things have moved on since then. Not only do we now have many groups of scientists engaged in ecological critique, but we also have much public debate spiced by the contribution from armies of eco-warriors. They run around pulling up fields of genetically modified crops and tunnel under projected road building sites. One recent commentator, typical of the mood in the West at the end of the century, announced, 'Goodbye to the oracle' (of science), and quoted the assistant director of the British Science Museum as saying, 'The days have just gone when experts could go ahead and make decisions without reference to the wider public' (Freedland 1999: 19). Politics may have become more and more scientised, but it is clear that today not only do the scientists represent environmental as well as industrial interests, but also it is clear that science itself has become politicised.

One British Sunday newspaper columnist even appears to suggest the very arrival of Habermas' rational society where science is mediated by citizens:

> Party politics no longer provides the ammunition for argument [at the dinner parties of the chattering classes]. Who argues about the minimum wage, New Deal or independence for the Bank of England?. ...The issues which divide us today are genetically modified foods, cloning, the use of embryo material for transplants, or the freezing of human eggs for later use by busy career women.

Most environmental issues – such as global warming or the dangers of air pollution – are based on scientific studies. Science is the new politics.

(Thomas 1998)

The sentiments in this quotation are no doubt subject to journalistic exaggeration. However, it seems that message of the basic environmental discourse concerning the threat that human activity, and particularly machine technology, poses to the planet's ecosystems has moved from being a fringe alternative discourse in the 1960s through being a dominant discourse following the 1987 Brundtland Report to being an embedded discourse by the turn of the century. This discourse demands constant evaluation of new technologies and, given that we cannot rely on industrial interest groups to carry out this evaluation, the demand for a more informed public discussion has grown. Even where governments defend the aims of the industrial groups, public discussion can ensure, as in the case of consumer refusal to buy genetically modified crops, that consumer pressure can achieve what pressure on government cannot achieve.

The actors who interpret and activate this discourse today are scientists and environmental NGOs. But what is the relationship between these groups?

Scientists and environmental groups

An impression is abroad that scientists are more influential than NGOs in identifying the existence and nature of environmental problems and that the role of environmental groups is to follow in their wake. This view is stated by Mazey and Richardson (1992: 121) when they say, 'environmental groups might be said to be one of the key links in modern society between science and politics, often being responsible for some kind of "megaphone" effect, transmitting scientific ideas from the private world of professional science into the world of public policy'.

There are perhaps two reasons for this impression. First is the demonstrable ability of scientists to deliver new information on various environmental problems; second, the dominance (despite the writings of Habermas and others) of positivist ideology in society at large, something that appears to be reproduced by Haas (1992a,b). In 1992 Haas introduced a series of articles on epistemic communities in the journal *International Organisation*, two of which were on environmental issues. Haas defines an epistemic community as 'a network of professionals with recognised expertise and competence in a particular domain and an authoritative claim to policy relevant knowledge within that domain or issue-area'. They have a shared set of principled and normative beliefs, shared causal beliefs involving multiple linkages involving possible policy actions and desired outcomes, shared notions of validity for weighing up knowledge and a common policy enterprise concerning the enhancement of 'human welfare' (1992a: 3). It is clear that Haas implicitly accepts an anthropocentric conception of environmentalism. Haas's ideas seem to be focused mostly on the international sphere.

Haas's article in *International Organisation* implies that epistemic communities can act in concert better than other transnational groups. Haas sees epistemic communities as having shared belief systems and consensual knowledge bases, things which he claims that social movements do not possess (1992a: 18). Scientists can allow the truth about the real world to shine forth in the end, he implies, even though he tempers his positivist position slightly by saying there is a role for a 'limited constructivism'. Although he concedes that, in providing consensual knowledge, epistemic communities do not necessarily generate truth, Haas says,

> The epistemological impossibility of confirming access to reality means that the group responsible for articulating the dimensions of reality has great social and political influence. It can identify and represent what is of public concern, particularly in cases in which the physical manifestations of a problem are themselves unclear, such as the case involving threats to the stratospheric ozone layer explored in this volume of articles.
>
> (1992a: 23)

Has Haas exaggerated the dominance of epistemic communities over environmental policy? Moreover, has this been done at the expense of de-emphasising the influence that environmental groups wield in shaping environmental policy?

The implicit claim that environmental groups do not have 'validity tests' of equivalent standing to those possessed by epistemic communities is questionable. Haas also implies that without 'experts' international environmental decision makers may 'ignore interlinkages with other issues but also highly discount the uncertain future, with the result that a policy choice made now might jeopardise future choices and threaten future generations' (1992a: 13). Just how important are the experts relative to environmental groups in making these linkages and steering governments towards goals involving ecological sustainability? The two environmental issues covered in the essays on epistemic communities introduced by Haas includes one, by Haas himself, about chlorofluorocarbons (CFCs) and ozone depletion, and another, by Peterson, about whaling. I will address both issues, starting with whaling.

Saving the whale

Peterson (1992) discusses the whaling issue. In the aftermath of World War II efforts were made by whaling nations under the aegis of the International Whaling Commission (IWC) to conserve whales for commercial purposes. The IWC was not especially successful in this endeavour and had little concern with promoting the ecological collective good of species preservation. Then, in the 1960s, environmental groups entered the debate. This built up into the 'Save the Whale' campaigns in the 1970s involving, for example, direct action by Greenpeace to engage the whalers on the high seas. In the early 1980s non-whaling nations were induced, under pressure from environmental groups, to join the IWC. This pressure culminated in a decision, made in 1982, for a commercial ban on whaling which came into force in 1986. Since then Norway and Japan have, to a limited extent, restarted whaling, although at much reduced levels compared to the period prior to 1986.

During this period, knowledge of whales began to be reconstructed. Their image changed from being dangerous sea monsters pursued by brave, if rather obsessed whalers, as pictured in *Moby Dick*, to sensitive, intelligent animals which were being killed in a cruel fashion by humans and were being driven towards extinction. Research into whale communication followed this new, increasingly dominant discourse, about the nature of whales, and discoveries were made about complex communications between the whales.

Peterson's argument illustrates how the cetologists (the epistemic community concerned with whaling) became increasingly fragmented as environmentalist arguments started to dominate the debate. In the mid to late 1970s, public interest was stimulated in the whaling issue 'as changes in ideas about the relation of nature and humankind began to affect attitudes on a broader range of questions' (Peterson 1992: 166).

Environmentalist pressure took off after the UN Stockholm Conference proposed a commercial ban on whaling in 1972. Conservationists launched the 'Save the Whale' campaign in 1973 and Greenpeace began direct action on the high seas in 1975 (Cherfas 1989: 209–12). It is significant that this heralded the start of the period 1974–82 (Peterson 1992: 170), containing, according to Peterson, the most serious divisions among the cetologists.

Haas, in his introduction to the series of epistemic community studies, implies that Peterson's work illustrates an example of the difference between epistemic communities and interest groups: 'If confronted with anomalies that undermined their causal beliefs, they [epistemic communities] would withdraw from policy debate, unlike interest groups' (Haas 1992a: 18). Yet Peterson's study does not demonstrate this. What it does demonstrate is that the epistemic community fractured along much the same lines as the contending interest groups (Peterson 1992: 154). Peterson tries to maintain that environmental groups differed from the epistemic community because they do not share 'causal beliefs, canons of validity, or principles' (p. 154), but this conclusion is reached only because Peterson apparently lumps all environmental groups under one heading before announcing that they are divided!

Analysts such as Robyn Eckersley have charted the differing strands of environmentalism (Eckersley 1992: 33–47), stretching from human-interest centred (anthropocentric) causes such as resource conservation and human welfare ecology, towards animal rights concerns and ecocentric orientations. Individual environmental groups tend to have their own ideological orientations which do constitute coherent sets of principles. Greenpeace, for example, prioritise ecocentric concerns that involve giving intrinsic value to nature (including whales) regardless of its usefulness to humanity, whereas the International Fund for Animal Welfare will concentrate on the cruelty aspect in keeping with their animal welfare orientation. Peterson's evidence suggests that epistemic communities tended to form around the norms set by particular interest groups rather than the other way around. This seems to explain why the cetological epistemic community fragmented in the wake of the increasing strength of environmental concern.

It seems that particular environmental groups, which promote norms based on particular world views, were far more principled and coherent compared to the cetologists' epistemic community, whose norms will be much shallower. This case study illustrates that epistemic communities can be more vulnerable to fragmentation than specific environmental groups. Environmental groups do not have the same canons of validity as epistemic communities simply because their principles do not require such tests. Logical positivists may challenge the meaningfulness of such principles, but then the debate has, at least in recent years, been largely shaped by ecological principles. When it comes to scientific issues, even Greenpeace spokespersons concede that the data on, for example, whale populations, are relatively uncertain. Thus debate will turn on interpretations of, and the degree of acceptance of, norms such as the precautionary principle rather than the scientific evidence *per se*.[3] The precautionary principle can even be used to question the necessity of having experts in environmental decision making (O'Riordan 1995).

New scientific information will inform environmental debates and may establish the existence of a problem. However, in cases such as the whaling dispute where humanity's relationship with nature is the fundamental issue, debates around the norms and the competing material strengths of the contending interest groups are the crucial factors determining outcomes. Environmental groups are in at least as good a position to form and project such norms compared to epistemic communities. Yearley has, through utilisation of a 'social problems' perspective, analysed how environmental groups shape and even sometimes 'invent' environmental issues (Yearley 1990: 47–77). But what about the second issue dealt with in the articles introduced by Haas, that of ozone depletion?

CFCs and ozone depletion

The CFC issue is different from that of whaling, in that there is at least broad agreement that the alleged affects of ozone depletion, caused by CFCs, are highly damaging compared to benefits derived from using the substances. It seemed that the only substantive issue was whether the theory of CFC-induced ozone depletion that was elaborated in the 1970s was borne out by physical measurements, which it was during the mid-1980s. Haas describes how the epistemic community which propounded the linkages involved in this issue helped develop and strengthen the 1987 Montreal Protocol.

In this case environmental groups did act much like a 'megaphone' for the epistemic community in, for example, driving CFC products out of the UK aerosol and fast-food packaging markets by means of boycotts (McCormick 1991: 105) or launching court actions against CFC producers in the United States (Seaver 1997: 65). However, the role of environmental groups was to extend well beyond merely acting as 'megaphones'.

A crucial factor behind the decision of chemical companies to accede to the growing pressure from scientists and environmentalists for action to phase out CFCs was the decision by these companies – Du Pont in particular – to look first for alternatives to CFCs that would do less damage to the ozone layer, and then to agree to produce these alternatives instead. But a key controversy, not covered by Haas in his piece in *International Organisation*, was about the choice of substitute. The chemical companies had a clear interest in choosing other, halocarbon, replacements for CFCs, such as hydrochlorofluorocarbons (HCFCs) and hydrofluorocarbons (HFCs) that they could patent and thus sell on to refrigerator manufacturers and other former CFC users at premium prices.

However, environmental groups including Greenpeace, protested that such substitutes were themselves damaging to the ozone layer and were also powerful 'greenhouse gases' leading to global warming. This was a policy problem that the scientists were unable to solve. First, there were different epistemic communities dealing with the CFC and the global warming issues (the latter now grouped around the Intergovernmental Panel on Climate Change), thus producing a problem of co-ordination and linkage. The CFC epistemic community wanted CFCs phased out, while the IPCC (formed in 1988) did not want them to be replaced by other chemicals that were powerful greenhouse gases. Greenpeace itself took on the task of investigating possibilities that HFCs were, indirectly, damaging to the ozone (Greenpeace 1994: 19). Second, there was controversy about the alternatives to HCFCs and HFCs. These are common hydrocarbons made from oil. Hydrocarbons have small potentials to increase global warming compared to halocarbons and do not damage the ozone layer, but for the chemical companies they have the commercial disadvantage of being unpatentable. The chemical companies thus faced the loss of a lucrative market if hydrocarbons were substituted for the halocarbons. In fact, in the 1930s, hydrocarbons like propane and iso-butane had been abandoned as coolants because of fears about their flammability. Chemical manufacturers ruled out use of hydrocarbons, and refrigerator manufacturers used the HCFC and HFC substitutes proffered by the chemical companies.

However, Greenpeace claimed that chemical companies exaggerated the problems with using hydrocarbons as refrigerator coolants. In 1991 they found that a small refrigerator manufacturer operating out of the former East Germany had foreign exchange problems in buying the HCFCs and HFCs from the chemical companies. Greenpeace funded the company so that it produced refrigerators using hydrocarbons as coolants. Since 1992 the use of hydrocarbons as coolants has spread, so that now all major German manufacturers and increasing numbers of other Continental refrigerator manufacturers use hydrocarbons as coolants. The notion that this is a fire hazard has been widely discredited, especially given that the amount of propane or butane actually used in refrigerators is very small. It seems that advances in manufacturing and handling these substances since the 1930s has largely eradicated safety problems.[4]

An environmental group was able not only to link together different environmental issues but also to sponsor research and engineering experience to demonstrate the efficacy of halocarbon-free refrigerators. It was better placed to do this than the scientists. Thus Haas's implication that epistemic communities have belief systems and consensual knowledge bases that environmental do not have is turned on its head. Greenpeace was clearly in a superior position to link issues and use their deep, normatively based belief systems to develop a halocarbon-free refrigerator strategy. Thus, even in the instance of the CFC issue, Haas's claims about the ability of epistemic communities to identify and represent what is of public concern (see the quotation on page 65) are exaggerated.

The above argument has concentrated in criticising Haas's conclusions about the alleged dominant role of epistemic communities in forming, articulating and co-ordinating ideas about environmental policy compared to environmental groups. But the ability of epistemic communities to produce value-free conclusions has also been attacked.

Beyond positivism

Liberatore has cited the controversy surrounding figures on different countries' shares of production of greenhouse gases to illustrate her contention that evidence and policy may be viewed differently by people from differing social contexts. India's Center for Science and the Environment have criticised figures produced by the influential Washington-based think tank, the World Resources Institute. The Center has claimed, for example, that carbon dioxide and methane emissions from developing countries have been overestimated. Liberatore concludes that 'scientists are socially situated reasoners rather than bearers of truth' (1995: 61).

As outlined earlier, Haas's work implies that epistemic communities are, because of their validity tests and commitment to apparent scientific truth, in a better position to judge environmental policy than environmental groups. Critics of the environmental movement such as Richard North put this in a rather more pejorative way by saying that environmental policy is much more responsibly handled by scientists rather than by environmentalists. He alleges that environmental groups are involved in a quasi-religious romantic crusade and that the rights or wrongs of causes they promote, such as animal rights or anti-roads protest, 'depend more on personal taste than scientific evidence' (North 1995: 41). I would agree with him on his latter point, although the judgement that environmentalists are mere romantics is highly contestable – see, for example, Andrew Dobson's discussion of the attitude of ecologists to the Enlightenment and the Romantic Movement (Dobson 1995: 10–12). But as the evidence in the cases of whaling, ozone depletion and global warming suggests, it is at least extremely difficult to find an environmental issue that does not depend on normative, socially constructed judgements beyond those that are generally accepted as being 'scientific'. Moreover, science is itself a social construct, given that particular truths are produced by society at different times for different purposes. I shall discuss the extent to which there are truth standards that are common to different human societies in Chapter 8, but whatever the outcome of this discussion there is no reason to suppose that scientists are any more able or qualified to make normative judgements than environmental groups.

The earlier discussion indicates that since the ecologically inclined scientists operate from the basis of the modern environmental discourse, any claim to absolute truth must be false. Their claim to act in concert with the environmental discourse may be legitimate, but discourses are not the possession of anyone, and environmental groups can also claim to act in concert with environmental discourses. As O'Riordan (1981) explains, environmental NGOs reject 'techno-fix' solutions ordained by experts. Instead, environmentalists see technical solutions as flowing from problems that are subjected first to normative judgements.

What is apparent in the work of Haas, and in a more tendentious sense in the case of North, is that they accept a broadly positivist position concerning the role of scientists as the legitimate bearers of truth, which, in view of the evidence, is unjustified. The position of all actors is shaped by normative factors springing from distinctive world views, although usually they seek to project their position as one that is neutral. One of the curiosities of the positivist claims made by Haas for epistemic communities is that they are the same claims as were made in the past by groups that defended unbridled industrialism and condemned those who suggested there were serious hazards being created by the industrial trajectory. The difference now, in the environmental sphere, is that the basic environmental discourse has become dominant, so the CFC epistemic community claims scientific objectivity on its side. Well, yes, but only in respect to the environmental discourse. Those defending a dominant discourse tend to lay claim to objectivity and condemn others who disagree with that discourse as being 'political'. This is the fate of radical environmental groups today whose own ideas develop the basic environmental discourse into versions that involve stronger notions of sustainability and often, the need for policies of social equity. But they are only 'political' in the sense that their ideas are outside the terms of dominant discourses.

The inadequacy of the claims to objectivity of the various protagonists in the eyes of many analysts and, intuitively, in the eyes of the wider public, means that we are moving into a position of 'post positivism'. This idea is complemented by Beck (1998: 161), who argues that scientific knowledge is no longer hegemonic in presenting evidence to an increasingly risk-sensitive public who may believe a 'science of NGOs' rather than official government or corporate science.

An epistemological position that accepts that there are many issues that cannot be verified by empirical observation and that policy prescriptions flow from the social construction of the issue will lead us to a model of environmental policy that allows us to analyse fully the importance of various interest groups in shaping outcomes. Environmental groups play a major role in this process which, in many issues at least, is much more than just acting as a megaphone for epistemic communities. Having said that, however, there is sometimes a considerable blurring of roles between environmental groups and scientists.

Blurred roles

Are the roles of scientists and environmental groups really so distinct? There is evidence to suggest that sometimes they are blurred in that environmental groups can sponsor scientific research and scientists can sometimes explicitly appeal to normative objectives.

In 1998, Greenpeace, as part of its campaign against radioactive discharges into the Irish Sea from the Sellafield nuclear reprocessing plant (based in Cumbria, north-west England), commissioned scientists to analyse samples of materials to look for radioactivity. In one case seaweed samples were picked up by Greenpeace employees and passed over to Southampton University for analysis. It was claimed that levels of technetium 99 had tripled in just a year (Greenpeace 1998a). Earlier in the year Greenpeace claimed that radioactive counts of pigeon flesh, feathers and faeces – contaminated soil it had recovered near Sellafield – were so high that the materials would be classified as nuclear waste. Caesium 137 made up most of the radioactive products detected, with some Cerium 144 also present. The analysis was conducted by the French-based Association for the Control of Radiation in the West (ACRO).

Greenpeace frequently commissions scientific analysis to support its campaigns. Critics say that this is really little more than propaganda, although environmentalists respond that spending by business on research to prove the benign nature of their products is far higher than anything environmentalists can muster. Nevertheless, there is no guarantee that Greenpeace's findings are without controversy. The organisation was seriously embarrassed in 1995 when it found it necessary to confess that it had wrongly declared that samples it had taken after it had boarded the Brent Spar oil rig in the North Sea showed clear evidence of pollution in the North Sea from the oil rig. Apparently they mistook the source of the samples that had been taken. Greenpeace had conducted a successful campaign to prevent Shell (acting with the support of the British Government) from disposing of the rig at sea. However, the fact that Greenpeace opposed the dumping before it found the alleged proof (on grounds other than the danger of pollution) does indicate that it was a case of looking for scientific evidence to back up a pre-conceived normative judgement rather than a normative judgement being made as a result of considering scientific evidence.

Some scientific efforts, like those of the IPCC and CFC epistemic communities, carry an implicit commitment to the environmentalist discourse. Others advertise it more explicitly, as their main objective. The EarthWatch organisation conducts research into flora and fauna conservation issues as widely based as conserving Indian wolves, Brazilian forests and the recovery of acidified water ecosystems in Bohemia. Its funds are raised mainly by private donation and its role is specifically orientated to 'saving the world's environment'. The EarthWatch Institute sponsors field studies 'to investigate and/or conserve the Earth's physical, biological and cultural heritage' (EarthWatch 1998). Is EarthWatch an environmental pressure group or an organisation of scientists? Perhaps both, but then its claims to scientific objectivity are certainly no worse than an organisation of scientists that is funded by ICI. The test that is increasingly applied to the scientists who are variously backed by environmentalists or business interests is how much their arguments and findings resonate with an increasingly democratised public debate. Complete democracy and transparency of information is, as yet, far from being achieved, but Habermas' 'ideal speech' situation seems less of a Utopia than it was twenty-five years ago.

We can see from this discussion that the emergence of the modern environmental discourse has coincided with a wider change in the relationship of science and politics, where the nature of technological progress has to be socially negotiated and cannot be left solely to industrial interest groups. This process, however, is not merely a matter of scientists who study the ecological ramifications of our technological path setting all of the agenda. The process also involves the making and shaping of belief systems and intrinsically normative judgements. This is where environmental groups play a key role. They can also, sometimes, play a key role in linking the concerns of different epistemic communities together and they can even sponsor scientific research themselves. What is very clear is that philosophical positivism is a poor basis for understanding the role of scientists and environmental groups, and indeed of the nature of environmental policy making in general. Scientists make implicit, and sometimes explicit normative judgements, and environmental groups perform an essential function in shaping belief systems so that we know what is important.

The approaches I have described so far – discourse theory, rational choice theory and an analysis of the development of, and agents connected with, the modern environmental discourse – do not represent solely detached academic analysis. The most prevalent politics associated with RCT is, via public choice theory, neo-liberalism. A critique of rational choice approaches to environmental policies and an elaboration of the discourse analytical approach to environmental policy can thus be expanded by a broader critique of the politics with which RCT is most popularly associated. It is to this that I now turn in a discussion of the discourses associated with neo-liberalism and green politics.

4
Neo-liberalism and Green Politics

It is now some 38 years since Galbraith (1999) announced that the West was enjoying an 'affluent society', yet despite real per capita GDP having much more than doubled since then, the West is now ever more competitively concerned with the objective of maximising economic growth. This period has seen the emergence of the modern environmental movement in the West, yet the movement's criticisms of the western obsession with economic growth appear to have had little impact. Although a mild version of environmentalism has been accepted into the western democratic consensus, it has only been accepted as an idea that affects the margins, not the centre of politics. Green parties are seen as 'single issue' parties by the majority and environmental pressure groups, although influential, are peripheral to the central economic debates.

Nevertheless it is still possible that a moderate green critique of economic growth could enter central economic debates. It could do so by questioning whether the present project of concentration on maximising economic growth as the all-consuming top priority, and especially the growing obsession with making society more and more competitive, is a way to maximise human, never mind wider ecological, welfare. Given the extent to which these moderate green questions are yet to be properly addressed it may be that green politics, or at least a version that more fully engages with mainstream, human-centred as well as nature-orientated politics, poses the only serious challenge to the now dominant neo-liberal version of competitive market capitalism. For let us remember that there is now a great vacuum in political economy. The notion of making economies ever more competitive is accepted as being axiomatic in countries such as the United States and the United Kingdom. By contrast, the notion of the planned economy and state control of key industries is seen as being an inferior method of achieving the materialistic goals of maximising economic growth that are common to the conventional left as well as the right.

Competitive market capitalism enjoins citizens, companies and public sector workers to ever more energetic forms of competition, with human capital increasingly talked of as another inert resource to be traded and evaluated as a marketable commodity. Performance-related pay schemes and short-term contracts are extended to wider and wider areas of employment.

Employees intuitively question this stressful turn of events, but they are, by and large, both persuaded and obliged to accept it in the face of the dominant, modernising discourse of competition. Even schools, which since public education began have been expected to instil virtues of caring and public co-operation into their pupils, are now increasingly expected to be driven by competition and to encourage their pupils to learn to compete in the market economy.

It may be that western publics, having surveyed the empirical evidence of the relative performances of market and planned economies in the light of the collapse of communism, have firmed up their support for the market model operating in the context of representative democracy. But this, in itself, does not explain the shift in dominant ideology, or, to use a term preferred here, 'dominant discourse', towards ever more competitive versions of capitalism since the mid-1970s, a shift that had already been well under way before the demolition of the Berlin Wall.

The function of this chapter is to describe and analyse two discourses: the neo-liberal discourse and the discourse of green politics. First, how did the aggressive version of competitive market capitalism that is commonly referred to as 'neo-liberalism' take hold?

Public choice and market theory

According to Self (1993), 'mainstream public choice theory has been fused with market theories and converted into a powerful new ideology that has become politically dominant over the last two decades' (1993: 56). In doing so this has dethroned previously dominant Keynesianism. It may be referred to as the 'neo-liberal discourse', and it is this which legitimises the ever more relentless drive for competition.

The discussion in Chapter 1 leads us to see a discourse as a set of meanings, categories, statements and ways of determining truths that form the basis for the generation of power and knowledge – power and

Neo-liberalism and Green Politics 77

knowledge being, according to Foucault, different sides of the same coin. Attention to discourse is both a method of analysis (through the examination of texts) and a technique for locating power/knowledge. For the moment I shall use Self's own analysis of public choice and market theory as a summary of the neo-liberal discourse, although in Chapter 6 I shall look at the example of recent developments in performance pay and education to look at how one key element of the neo-liberal discourse, the idea of competition, is being applied.

Public choice theory asserts that interest groups and bureaucrats will inevitably manipulate the government system to feed their own purposes and that the political market has an inherent tendency to oversupply public goods. Public choice theory takes its cue from economics, including the idea that consumers have a rational ordering of preferences and that they pursue their self-interest in accordance with this preference structure. Public choice's theory of rational preferences is in line with that of market theory, except that in public choice theory the preferences are about political and social matters rather than economic issues.

In fact, a careful reading of Olson and other public choice theorists reveals that the explicit emphasis on maximising competition that forms such a key part of neo-liberalism is already implicit in public choice theory. Olson discusses how, in a perfectly competitive market, firms have a common interest in keeping prices high, but no common interest when it comes to output where the interests of the individual firm lie in maximising their output. Under perfect competition prices and profits fall to their lowest level, apparently the most economically efficient result, but of course contrary to the individual interests of the firms. Crucially, Olson comments that 'About the only thing that keeps prices from falling in accordance with the process just described in perfectly competitive markets is outside intervention' (1965: 10). It is not hard to see the path of Olson's logic, which moves from seeing how economic markets are distorted by intervention to seeing how political decision making is distorted by small groups who can feather their own nests with relatively minimal political effort. The implication is that the distribution of goods and services needs to be left to the market, and a market which is as perfectly competitive as possible. Most analysts would accept the economic theory of perfect competition, assuming that such perfect competition is possible in a given situation, but many would dispute the notion that the simple test of achieving the lowest prices can be the basis for maximising welfare throughout society. This becomes especially clear when the concept of external costs – costs that accrue to society but which the agent causing those costs does not prevent or pay for – are taken into account.

Despite this, public choice theory has been imported to political science in the context of the fusion of the strong version of public choice theory with market theory. The resulting 'neo-liberal' discourse 'assumes that market systems are inhabited and operated by rational egoists and that under competitive conditions the results will be generally beneficial' (Self 1993: 198). This neo-liberal discourse also takes on board the notion that the role of the state should be restricted to creating the conditions suitable for the smooth operation of free markets. In addition, the discourse asserts that state intervention in economic affairs will not act to improve social welfare, but will give opportunities for self-interested groups to gain favours and for public sector bureaucrats to expand their fiefdoms.

I am most concerned here to discuss and analyse the implications of the emphasis on market-based competition that is projected by neo-liberalism. As we shall see, this has great implications beyond any gains in economic efficiency that may, or may not, be the result of the increasing emphasis placed on making society, and the organisations and people within it, more competitive. Although analytically distinct, public choice theory and classical market theory are used by the same analysts, including Buchanan, Niskanen, Olson and Friedman, who have been associated with new right economic and political thinking. These theorists assume that the pursuit of material self-interest in politics is universal (Buchanan 1986). Analytically, of course, public choice theory is part of rational choice theory, which I examined in Chapter 2 in relation to environmental policy. However, public choice theory has its own distinctive set of normative implications.

The norms of public choice

That public choice and neo-liberal theorists have a normative agenda is often masked by appeals to apparently objective (positivist) conclusions of economic theory. However, some theorists do not hide their moral standpoints. Authors like Nozick and Hayek argue that redistribution of income and wealth has no moral basis is an affront to the personal freedom to dispose of one's income as one wishes. In addition they promote the belief that, as Gamble puts it: 'Any attempt [by the state] to enforce a particular pattern of distribution must destroy many of the benefits that flow from an unhampered market' (1994: 61). Friedman describes the 'capitalist ethic' whereby:

> Payment in accordance with product is therefore necessary in order that resources be used most effectively.... No society can be stable unless there is a basic core of value judgements that are unthinkingly accepted by the great bulk of its members. ... I believe that payment in accordance with product has been, and, in large measure, still is, one of the accepted value judgements.
>
> (1982: 166–7)

The fact that the path to income redistribution advocated by the left has been through the state and that the right have advised a reduction in state intervention takes attention away from what could be interpreted as a sleight of hand in the arguments of the new right. By arguing against state intervention they appear to take it for granted that efforts to reduce inequality always require state intervention and that the 'payment in accordance with product' involves no moral judgement since this is decided by the market which, it is assumed, involves no moral judgements itself. However, the market is a confluence of moral, normative judgements made by consumers about what they think it is important to purchase, whether it be deciding which car suits their image or how important it is to hire somebody with a particular skill as opposed to somebody with another skill. This downplaying of the role of social norms is one of the various analytical traits of public choice theory which it shares with rational choice theory, an issue that has been discussed in Chapter 2.

The central question here is, how far can we move towards equality of income? We may assume that perfect equality is not possible nor even desirable, but the degree of inequality that is acceptable is a moral judgement that is a very real one within a given organisation, whether private or public. Individual organisations, of course, will be very much affected by prevailing social norms on this subject, which have tended over the last 20 or more years to regard growing income inequality as more acceptable than it was in the past. The trend towards greater income inequality is reflected in the way tax systems in the West have become less progressive over the last 20 years. However, another issue is what people are paid before tax is taken into account. Judgements about the relative worth (in pay terms) of people with differing skills are governed by social norms, not some transparently available body of knowledge about what skills are most useful to companies and public institutions. Is the relative pay of accountants and engineers decided by sheer notions of economic efficiency, or is it governed by changing social norms about the relative worth of these professions, norms which vary from country to country? In the United Kingdom, for example, there is a continued complaint that accountants are valued by companies more than engineers. How is it that the difference between the pay of senior executives and other employees has widened considerable since the 1970s? Has this something to do with the changing usefulness of chief executive officers, or is it more to do with the influence of the neo-liberal discourse which implies that there is nothing wrong with relatively high levels of income inequality?

The normative nature of public choice theory becomes very apparent when one applies the theory to the private sector pay awards of senior executives. They decide on one another's pay through remuneration committees, and are thus in a privileged position compared to the shareholders. Will nobody apply for senior executive positions if the general level of salaries for such posts stops their seemingly exponential rise? I doubt it, but perhaps the social constraints on the increasing tendency towards boardroom pay excess have been weakened by the implicit normative agenda of the neo-liberal discourse.

The left's traditional commitment to equality has been considerably weakened, since the 1970s, by the neo-liberal discourse. This is a discourse that even centre-left governments of the Clinton and Blair type had absorbed by the time they took office. As we shall see later, the Blair Government developed a preference, among other things, for encouraging competition among public sector workers so as to maximise output. We can see the justification for the emergence of short-term contracts, the consequent erosion of job security and the introduction of performance pay-related schemes in the terms of neo-liberal discourse. Job security is seen as a constraint, an outside intervention in the market, and the matching of rewards with productivity makes economic activity more competitive. But do such stratagems, which seek to produce goods and services via the lowest possible price, maximise happiness? Could it even be that the very ultra-competitive methods used to produce the lowest price end up undermining this objective?

The neo-liberal discourse has been linked to a discourse on globalisation which implies that we have moved towards a globalised market of competing goods and services. However as Watson (1999: 131–2) points out, while there may have been a globalisation of financial markets, there is little sign of a globalised market in the trade of goods and services. In the case of the major economies 90 per cent of goods and services continue to be produced and consumed by the domestic markets. Hence the central justification for new-liberalism's emphasis on super-competitiveness and the need to drive down production costs is based on a myth.

I am most concerned to tackle the implications of the neo-liberal discourse for green politics and environmental policy. This is likely to throw some light on the possibilities for green politics to provide an alternative to this discourse, a discourse that forms the intellectual driving force behind the ever quickening pace by which our economies are ever more deeply and our social life is ever more widely subjected to the pressures of competition through the market.

As I implied before, the success of this theory cannot be put down to the political victory of capitalism over communism, for the neo-liberal discourse had gained hegemony already during the Thatcher–Reagan period. So how has it become predominant?

Middle classes and material self-interest

In the 1950s and 1960s social democratic explanations, such as that put forward by Habermas, for the survival of capitalism hinged on the idea that welfare programmes had been established, paid for out of taxation, so as 'to compensate for the dysfunctions of free exchange' (Habermas 1971: 102). Hence European social democrats and American liberals thought themselves able to pursue goals of equality by expanding the proportion of GDP devoted to state spending. But by the 1980s such analyses and goals had apparently become less convincing in the face of the brakes that had been placed on expansion of the welfare state. These brakes were applied on ideological grounds by right wing governments and applied by centre-left governments on the more pragmatic grounds of pacifying voting electorates that had become less inclined to pay higher taxes, especially higher taxes on income.

According to Self, the public choice market theory (what I call the neo-liberal discourse) was spread by right-wing think tanks. Although this may describe the way in which this discourse was diffused among conservative politicians, it does not explain why this discourse achieved the resonance it did in order to achieve dominance, or hegemony.

However, Galbraith talks of a plausible explanation. He tells of the growing size of the contented and relatively wealthy middle classes, classes that feel the need for the state as social insurance less strongly than many of their parents. Thus, for example, the contented resent the need to pay taxes to support what has become known as the underclass. Political parties which promise tax cuts now find an appeal that is much wider than was the case during the days when the 'contented', or at least those who could realistically aspire to be contented, were much smaller in number.

> What is new in the so-called capitalist countries... is that the controlling contentment and resulting belief is now that of the many, not just of the few. It operates under the compelling cover of democracy, albeit a democracy not of all citizens but of those who, in defence of their social and economic advantage, actually go to the polls. The result is government that is accommodated not to reality or to common need but to the beliefs of the contented, who are now the majority of those who vote. A consensus old as democratic government itself still prevails.
>
> (Galbraith 1992: 10)

Galbraith talks of 'the majority of those who vote' because voter turnouts are significantly lower among economically marginalised groups, a point which is relevant to my discussion, in Chapter 5, of the negative consequences of social division that do not appear directly in accounts of economic transactions.

Galbraith explains that the doctrine that is needed to legitimise the policies favouring the contented needs to fulfil three requirements. First, it 'must be a doctrine that offers that offers a feasible presumption against government intervention', second, it needs 'to find social justification for the untrammelled, uninhibited pursuit and possession of wealth', third it needs 'to justify a reduced sense of public responsibility for the poor' (1992: 96–7).

The doctrine that performs these functions is the neo-liberal discourse that has been discussed earlier. Galbraith points out that policies that are justified by this doctrine, such as higher interest rates, or higher pay to corporate leaders, are also ones that happen to benefit the most contented members of society. However, is this apparently straightforward account of the application of material self-interest all there is to say?

Interpreting self-interest

Put this way, the changes brought about first by enfranchised workers seeing their self-interests being maximised by increasing public spending on social welfare, and later by a growing majority of voters who see themselves as middle class with less need of the state and high levels of taxation, represent an example of the explanatory ability of RCT in general. RCT, as applied to political science, is a methodology that is concerned with the analysis of how individuals and institutions seek to maximise their own self-interest. The now middle-class majority of voters have plumped for the arrangements that maximise their perceived material self-interests. However, this use of RCT is dependent on self-interest being interpreted as maximising self-interest in purely financial ways. The same can be said for public choice theory except that public choice theory makes specific (and controversial) claims about how self-interest affects the role of the state which have helped to give legitimacy to the currently dominant neo-liberal discourse that says that more and more competition is good.

Marxists and many of these adhering to traditional left-wing thought have been hamstrung by the growth of this contented majority that has precisely reversed Marx's predictions of increasing immiserisation of the majority. The problem with traditional left-wing politics is that, for a great part, it was, and remains, based on the same calculations of individual material self-interest as the economic liberals who have continued in the tradition of Adam Smith. Rational choice theorists might be expected to agree that the left's most potent appeal in electoral terms was not that socialism was morally superior, but that it was the best way of satisfying the material aspirations of the majority. In this interpretation, class politics are no more than an aggregation of the interests of individuals who are said to occupy a similar social location.

However, two criticisms of this aforementioned rational choice analysis might be made. First, most would agree that the feeling of social solidarity engendered during World War II provided an important moral underpinning of the social democratic, post-war consensus in the United Kingdom, thus helping to shape perceived notions of self-interests to favour a larger public sector (Addison 1975). Second, since the early nineteenth century, there has been a significant strand of moral socialism including, for example, the Co-operative Movement (Cole 1944). Under moral socialism – or even the collectivist capitalism that exists in Japan – a degree of equality is something to be pursued because it is beneficial to the functioning of the whole of society, not merely because it is in the individual interests of the members of the working class.

However, as the memory of World War II faded, so the post-war consensus was evaluated less in terms of the equality that was produced and more on the basis of perceived individual material interest. As far as the left is concerned, the existence of moral socialism has generally been overshadowed by the material concerns of trade unions whose chief preoccupation has been wage increases, the activities of Fabians whose justifications for state intervention tended to be based on efficiency, and Marxists, who followed Marx in denouncing moral socialists as Utopian.

Socialism has, for good or ill, been seen to fail in terms of the material objectives which the key sections of the left held to be so important. Economies with extensive public ownership, let alone centrally planned economies, are now out of favour. The dominant discourse has shifted away from mixed economies to talk of making competitive market economies even more competitive and in creating markets where before none had even been imagined. Part of the logic upon which this is based is that so long as they do not actually make many people poorer, super-competitive economies are bound to deliver higher economic welfare. Increasing income differentials is interpreted not as signs of moral decay but as incentives to do better. The neo-liberal discourse reigns supreme, at least in the United States and the United Kingdom. By contrast, left-wing discourses are seen to be more inefficient means of obtaining the same material objectives to which the neo-liberal discourse is committed.

But the neo-liberal discourse is interesting not just because it merely reflects the perceived material interests of the growing middle classes, but also because the discourse itself may be highly significant in shaping people's aspirations, motivations and attitudes to public policy. In this case (that is, attitudes to competition), as in other instances that I shall discuss later, discourse not only reflects perceived self-interest but it also acts to shape the interpretation of that self-interest. People may think that more competition is good because it is sanctioned by a discourse that resonates with their preferred political programmes; for example, tax cutting. They may thus support political programmes and policies that put in place more competition and spurn approaches and ways of doing things that rely on co-operative and egalitarian norms. Moreover, they may be more likely themselves to act more competitively and interpret their own and other people's degree of success in terms of how successful they have been in the competitive exercise. Status is determined by the degree of competitive success, and success, in the terms laid down by the neo-liberal discourse, is associated with maximising material rewards.

I have suggested that the neo-liberal discourse has shaped peoples identities and interpreted self-interest as being concerned with the pursuit of material self-interest through competitive means. If there is a discourse that can challenge the supremacy of this super-competitive version of market capitalism, it has to be a discourse that reinterprets self-interest differently. What about some discourse that promotes altruistic objectives? Public choice theorists such as Olson dismiss the notion that people will act altruistically. Indeed, altruistic or 'emotional' political action was described by Olson as 'irrational' (1965: 108–9). It may indeed be unrealistic for altruistic action to be the basis of society (although, as we have seen in Chapter 2, public choice theory does not explain altruistic campaigning by environmentalists), but it is still possible for people to interpret their self-interest differently from that defined by the neo-liberal discourse. For example, people can interpret their self-interests as being associated with support for public policies that depend less on competition. However, it seems unlikely that traditional left-wing politics can provide the discourse that can reinterpret people's self-interest, for left-wing politics, as discussed earlier, are themselves concerned with prioritising the same materialist ends as the neo-liberal discourse. It is just that their means have proved unpopular with the expanding middle classes. Now I want to investigate the extent to which green politics offers a platform from which to launch the reinterpretation of self-interest away from the concentration of competition that is so evident today and which has been legitimised by the neo-liberal discourse.

Some would regard this juxtaposition of green politics with neo-liberalism as somewhat odd. Surely, if you are going to discuss neo-liberalism and environmental policy you ought to be talking about whether market instruments, green taxes and other practices associated with neo-liberal market approaches are compatible with achieving environmental objectives, they may say. Indeed, this might be the correct thing to do if I were just talking about environmental policy, and I shall certainly take up these questions later on when I analyse some key aspects of environmental policy. But to do so now would be to conflate green politics and environmental policy. The latter is indeed addressed by the former, but only in the same sense that there is an environmental policy subset (or at least should be) of neo-liberal theory. In other words, green politics is much more than the environment – or at least has the potential to be so even if its initial, and so far its most central, debates have hinged on the relationship between humans and external nature.

Can green politics reinterpret self-interest?

What is widely considered to be the first wave of environmental consciousness occurred in the latter part of the nineteenth century when environmental groups were established principally to protect what were seen as aesthetic countryside and (in the case of the United States) wilderness interests, although 'preservation' for its own sake was a feature of the emergent environmental movement in the United States. It was in the 1960s that the modern environmental movement formed, with its concern that pollution and resource depletion were threatening human ability to survive. In the 1980s a third wave emerged, carrying forward 1960s themes with a more explicitly global agenda and a switch in concentration away from resource and population problems and towards pollution. I discussed the development of environmental ideas as a discourse in Chapter 3. What concerns me now are the philosophical trends that have developed.

Unlike the left, green politics has featured a critique of economic growth at its heart. Broadly speaking, greens argue that during the course of production of material goods, external costs – costs that are not reflected in the commercial transactions – are inflicted on society and the environment through environmental degradation. Disagreements exist among greens concerning the extent to which economic growth can be reconciled with sustainability, the project of continuing human activities that will not damage the ability of future generations of humans, and possibly other species, to flourish. There is debate on whether, and the extent to which, consideration of non-humans is necessitated by the sustainability concept.

Could green politics, one not dogmatically opposed to all development, yet be the basis of a more profound reinterpretation of currently materialistic and competitive notions of self-interest? Although it is not the function of this book to engage in a full analysis of green political philosophy, it is necessary to sketch in some key points of green philosophical debate if this question is to be answered.

As Vincent (1992) has commented, there is a disjuncture between the political practice of the environmental movement, which emphasises human prudential reasons for action, and the thoughts of eco-philosophers, who have since the early 1970s developed a tendency to base ecological principles on the defence of the intrinsic value of nature. Vincent identifies this as a weakness and describes it as 'strange' for a political movement (1992: 236).

It is since the early 1970s that many eco-philosophers have placed increasing importance on the value of nature independent of human interests. Naess, in a lecture delivered at a Futures Conference in 1972, coined the distinction between nature-centred deep ecology and shallow ecology, concerned only with narrow human interests. He implied that humans must develop a close relation with nature, and promoted the concept of 'biospherical egalitarianism', meaning that all forms of life have 'an equal right to blossom' (1973: 96). This was an introduction for the concept of ecocentrism, with Eckersley defining the term as involving upholding the intrinsic value of living things. Indeed, Eckersley describes a continuum of ecological views ranging from the most anthropocentric trend of being interested in conserving nature for resource efficiency purposes, to radical human welfare ecology, preservationism, animal rights (which cares for individual animals rather than species) to the most radical trend, ecocentrism (Eckersley 1992). Deep ecology, sometimes known as 'transpersonal ecology', has latterly been developed by Devall and Sessions and others to involve a human identification with a wider life-force 'self' (1985).

Dobson (1995), who covers much of the same ground as Eckersley, adds a distinction between environmentalism (which he sees as a reformist movement capable of being absorbed by mainstream politics) and 'ecologism' (which is distinct on account of its support for ecocentrism, opposition to economic growth and preference for decentralised political arrangements).

There are several grounds for criticising this wish to define green politics as being identified with defending nature on the basis of its having intrinsic value or, in later deep ecology works, as being concerned with the development of a metaphysical identification with nature. Quite apart from the existence of accounts of green politics such as Paehlke (1989) and Goodin (1992) that are coherent without relying on the anthropocentric/ecocentric distinction, the term ecocentric has been defined, earlier by O'Riordan (1981), in terms that do not involve a notion of valuing nature independently of human interests. O'Riordan, in contrasting his ecocentric idea with mainstream 'technocentric', quick fix, end-of-pipe solutions to environmental problems, suggests that humans need to respect nature since they are subservient to the force of nature. Norton (1991) cites Aldo Leopold's 'Land Ethic', which rather like O'Riordan, requires humans to take those actions which respect the integrity of the land in order to protect human as well as natural interests. Hayward attempts to bridge the gap between the deep ecologist position and O'Riordan's 'enlightened' human self-interest position by suggesting that, while only humans are autonomous agents and therefore able to make ethical judgements, humans should have both rights and obligations including the obligation to pursue the realisation of human autonomy without doing so at the expense of non-humans (Hayward 1994: 84).

Such approaches threaten to make the appeal to the intrinsic value of nature irrelevant. Barry (1999) has commented that

> [S]ince anthropocentrism has not been demonstrated to be either bankrupt or a 'dangerous' orientation to the non-human world (Fox 1990), the compulsion to search beyond anthropocentrism for an appropriate moral idiom loses much of its force. An immanent critique of anthropocentrism ought therefore to be the strategy adopted in order to achieve public support for the normative ends of green politics.
>
> (1999: 41–2)

Barry puts forward a coherent philosophical and practical framework to support his argument that humans should perform a role of ecological stewardship (in both a rural and an urban setting), and that humans should settle ecological arguments by reference to how humans relate to nature in terms of practical debates about specific issues. He says that the evocation of emotions such as sympathy are a surer basis for humans agreeing to set aside inessential interests in order to avoid harm to non-human beings than metaphysically based ethics. His arguments tend to convince one that the solution to the disjuncture raised by Vincent is for the ecophilosophers to change their position so that they are more in concert with the practical environmentalists who try to speak in language which most people understand.

There is a profound problem with attempting to underpin ecological arguments on deep ecological ethics. Although human impact on nature is disruptive, leading (for example) to an acceleration in species extinctions, it is still the case that human impact on the environment poses a much greater threat to human existence than any threat posed by humans to the existence of life on the planet. Lovelock comments:

> From the Gaian point of view, the evolution of the environment is characterised by periods of stasis punctuated by abrupt and sudden change. The environment has never been so uncomfortable as to threaten the extinction of life on Earth, but during those abrupt changes the resident species suffered catastrophe whose scale was such as to make a total nuclear war seem, by comparison, as trivial as a summer breeze is to a hurricane. We are ourselves a product of one such catastrophe. Could it be that we are unwittingly precipitating

another punctuation that will alter the environment to suit our successors?

(1988: 153–4)

There is yet a further way in which the deep ecological attempts to reinterpret human interests (or more correctly to displace them in favour of a nature-centred altruism) is misconstrued. In the process of trying to move humans closer to an external nature, they are in fact moving ecological attention from the ways in which humans are holistic beings in the sense that, for example (as we shall see in the case of stress and low status in the following chapter), the internal psychology and physiology of humans can interact to produce unhealthy human outcomes. The competitive market economy can produce increases in the stream of material goods, but as we shall see there are strong arguments that there are external costs to society; for example, damage to human health and an increase in violence, caused both directly through stress as well as indirectly through pollution. Just as nature should not be seen as some sort of machine, so greens should not see humans as no more than instrumentally motivated (ends-related) machines whose actions have to be shaped to fit in with nature. These themes will be developed further in succeeding chapters.

There has been little appreciation of this, in a practical sense at least, within green political theory. John Barry goes some way towards it when he argues that 'The reconciliation of green philosophy and politics depends on seeing that the normative basis of green politics includes a concern with the social world and its organisation, as much as moral concern with the non-human world' (1994: 369). However, Barry does not discuss the issues raised in the previous paragraph. His analysis of green political economy provides a solid basis for the political and economic aims of conventional green politics and the major social and political restructuring that this demands. However, it is justified mainly for the benefits it secures for nature and indirect human benefits through pollution reduction.

Barry identifies four ways in which green political economy differs from other versions of contemporary political economy. First, it gives a special place to institutions such as regimes governing the commons in the regulation of the economy. Second, it questions the ends of development and it criticises the dominant attitude to consumption. Green political economy sees work and production processes as more than just instrumental means of productionist ends. Third, it questions the notion of what is a resource and challenges the notion that environmental resources are effectively infinite. A relationship between environment and ecology has to be negotiated on the basis of the sustainability principle. Fourth, green political economy has a much wider notion of what is meant by economy compared to other versions of political economy. 'Consumption needs to be integrated with a more expansive mode of acting and experience which recognises the dependence of human productive activity on the creation and maintenance of a stable metabolism with nature' (Barry 1999: 184–6).

I would not dispute these sentiments, but his call for economic analysis to be shifted 'away from a consumption-driven and consumption-centred economy' (1999: 185–6) should be founded not only on ecological sustainability criteria but also on wider, social sustainability criteria, the reasoning for which will become clearer in the succeeding chapters of this work.

The perceived failure of green political movements to deal with mainstream concerns has relegated them to being tagged as 'single-issue' movements. In many ways this is unfair, especially considering the importance of developing ecologically sustainable policies, but it is a criticism that can be dealt with if greens can extend their critique to directly human-based concerns and do not isolate themselves by defining green politics purely in terms of defending nature. After all, greens do propose changes to human society that have far wider ramifications than ecological concerns.

Broadly speaking, greens promote what seems these days, in an era of neo-liberal dominance, to be concerns for social justice that are shared with social democrats, albeit justified by greens on the basis of the need to make equal sacrifices in the face of limited resources. Although Jonathan Porritt opposed the positioning of green politics on the left, his major work espouses co-operative values and methods, non-materialist outlooks and social equality (Porritt 1984). What greens, including Porritt, all too often fail to do, but what there is great potential to do, is to justify such strategies on the basis of a green moral discourse that adopts a holistic attitude towards relationships between humans as well as between humans and nature. Interestingly, the social agenda supported by Porritt in 1984 is now, because of the establishment of the Thatcherite/Reaganite neo-liberal consensus and the relative rightward shift of political attitudes, quite definitely on the left of today's Anglo-Saxon politics. This presents an excellent political opportunity for green political economy to colonise a vast region previously occupied by the conventional left. A green justification for social equality that includes a priority for mainstream economic and social concerns in their own right is now potentially much more credible than the old left's contention that their systems of central planning are more efficient than those operating under capitalism. What could be called a 'social green' approach will be discussed in Chapter 7.

This argument should not be read as suggesting that green politics is a subset of social democratic politics. On the contrary, what it does suggest is that social democratic aims of social justice, although not social democracy's traditional centralising tendencies, can be seen as a subset of green politics. The green critique, with its concept of the external costs of production, can be adapted to look at the human costs of the competitive market economy. Used this way the green analysis of external costs can incorporate both concerns for the environment and concerns for the impact of social division and excessive competition. If there are significant external costs of excessive competition in pursuit of the goal of material production and consumption, then the green critique could indeed become more potent. Thus I shall now discuss some of the evidence for the existence of external costs of increasingly institutionalised, excessive competition that are visited directly on the human as well as the non-human world. We can begin this next stage of the investigation by looking at perhaps the most fundamental issue of all, the relationship between material well-being and health.

5
Health and Materialism

In most societies there can be observed a general improvement in health, as measured by life expectancy, as social and scientific learning about improving health is refined. This may not be a perfect measure of health. For example, longer life expectancy does not always involve an extension of high-quality life. Nevertheless, life expectancy is a concept that is easily measurable and about which there is little argument over interpretation. To that extent it is a useful guide to the relative state of health in different societies at different times.

Although the general upward improvement of life expectancy in most of the world is clear, the relationship between given levels of material living standards and health standards is not so straightforward. Some countries with living standards that are only a fraction of others still have as long, or possibly even longer life expectancies. Indeed, beyond a certain threshold of guaranteeing hygienic water and sewerage services and care for children, increases in material living standards do not necessarily lead to any improvement in life expectancy at all. The 'threshold' can be associated with surprisingly low material standards, as measured by per capita GDP, as can be seen in Table 5.1. This table consists of some selected countries with differing per capita GDPs. I do not claim that this table is particularly representative – although I doubt if it is wholly unrepresentative, but it does demonstrate the lack of clear links between GDP and life expectancy.

Countries with per capita GDPs of only a tenth of those of the USA (that is, less than what the average US citizen spends on health care alone), such as Costa Rica, manage to achieve a life expectancy that is similar to that of the United States. Given the high medical expenditures in the 'advanced' industrialised states, one begins to wonder not just about whether all the money is well spent – but what other

Table 5.1 GDP and life expectancy in selected countries (1993)

Country	Per capita GDP (1994 US$)	Life expectancy (1993)
USA	25,900	75.9
UK	17,600	76.1
Spain	12,300	77.1
Cuba	3,500	75.4
Costa Rica	2,600	76.3
Maldives	530	62.1
Nigeria	340	50.5

Sources: United Nations Department for Economic and Social Information and Policy Analysis, Statistics Division (1996); *UN Statistical Yearbook – 41st Issue*, New York: United Nations and Euromonitor plc (1996); *The World Economic Factbook*, (1996) Chippenham, Wilts: Anthony Rowe.

countervailing factors appear to neutralise this apparent advantage when comparisons are made with other countries that are very much poorer. What have the poorer countries got, or not got, that the rich countries have? The weak correlation between GDP and life expectancy that is evident when comparisons are made between different countries that have a basic minimum of services represents an anomaly for the dominant neo-liberal discourse, which holds that the maximising of material living standards is the most important social objective. Of course, in much the same way, it also represents an anomaly for the conventional left.

When people are confronted with the life expectancy–per capita GDP anomaly they often dismiss it by saying that people in richer countries eat the wrong foods or, sometimes in an implicitly racist commentary, argue that, for example, the life expectancy of US citizens is weighed down by poor health among non-whites. Whereas it is indeed true that the life expectancy of a black male living in New York is, at 56, no more than that of the average citizen of Bangladesh (one of the poorest countries of all), countries like Cuba and Costa Rica consist precisely of the ethnic minorities that are supposed to reduce US life expectancy. This suggests that the causes of the life expectancy–per capita GDP anomaly have something to do with the way countries like the United Kingdom and the United States, as advanced capitalist societies, are organised, not racial characteristics.

Let us now turn to some of the evidence to suggest that inequalities in health and some other social problems are associated with inequalities of income.

Inequality, health and social malaise

Of course it is the relatively deprived who suffer low status, and, as we shall see later, greater stress. Income inequalities are also indicative of major differences in life expectancies. According to a study conducted in the Netherlands, there is a 12-year difference in life expectancy between the highest and lowest status individuals, as measured by educational achievement. In fact these differences are smaller than those existing in the majority of other western states, including Britain, France and the United States, the latter having the biggest disparity in mortality rates between high and low educational achievers (Mackenbach 1994: 1488).

Such differences cannot be ascribed mainly to material differences, for even the best-educated persons in much poorer countries like Costa Rica and Cuba earn no more, and, depending on arguments about the buying power of currency at a local level, perhaps rather less than those on relatively very low incomes in countries like the United States and the United Kingdom. Yet the 'poorer' citizens in Cuba and Costa Rica live much longer lives than the 'richer', but lower status, people in the United Kingdom or the United States.

The research suggests that not only inequalities in income, but also in health as measured by life expectancy, have grown in the United Kingdom in the 1979–97 period, precisely the period in which policies associated with the public choice–market discourse have been put into practice (Acheson 1998).

In general, such growing inequalities of health have occurred in the context of an overall improvement in life expectancy that has been ongoing for the last century, even if this improvement has latterly been delivered disproportionately between different social categories. However, there are circumstances where even this trend has been reversed, most graphically in the case of Russia, which has attempted to engage in a rapid embrace of free market economic strategies. In the period 1990–94 average life expectancy actually fell by five years. Researchers have noted how this has been accompanied by a decline in social cohesion and a tremendous increase in crime, with the sharpest increases in crime coinciding with those areas where there has been the sharpest decline in life expectancy (Walberg *et al.* 1998). Some may say that what has happened in Russia is but a particularly extreme and unfortunate example of what has been happening in western states such as the United States and the United Kingdom since the 1970s.

There has been an extensive body of medical literature built up in recent years linking declines in income equality with growing disparities in health in many countries. Richard Wilkinson has been a leading researcher in this field. In an early work (1992), in which he studied health and income inequality in various industrialised states, he suggested that average life expectancy in the United Kingdom could be increased by two years if income equality was brought up to average European levels. Wilkinson also pointed out that the country with the highest life expectancy – Japan (79.5 years in the 1990–95 period) – also had the greatest amount of income equality among the industrialised nations.

Wilkinson assembled various studies to demonstrate the relationship between income inequality and life expectancy in developed countries. For example, one study (Bishop *et al.* 1989) shows a high (Pearson correlation 0.86) correlation between life expectancy at birth and and percentage of post-tax and benefit income received by the least well off 70 per cent of families. There were nine countries in the study. West Germany, the United Kingdom, the United States and Australia were the four countries with the lowest life expectancy. They had a combined average life expectancy close to 74 years and an average of 46 per cent of income received by the least well-off 70 per cent of families. The four countries with the highest life expectancy – Norway, Sweden, the Netherlands and Switzerland – had an average life expectancy close to 76 years and had an average of around 49 per cent of income received by the less well-off 70 per cent of families. Canada was in between both groups on both life expectancy and income equality measures. In other words, life expectancy shows a fairly strong association with the degree of income equality.

An inquiry commissioned by the UK Department of Health cited some of Wilkinson's later work (and work by criminologists) as evidence for making the judgement that

> Recent research suggests that, in addition to the ill effects due to absolute poverty, societies in which there is a wide gap between the rich and the poor suffer additional social problems, for instance, through high rates of violence and crime, and truancy. It has also been suggested that people with good social networks live longer, are at reduced risk of coronary heart disease, are less likely to report being depressed, or to suffer a recurrence of cancer and are less susceptible to infectious illness than those with poor networks.
>
> (Acheson 1998: 9)

The key issue here is that income inequality, and by implication the move towards a more competitive society as prescribed by the neo-liberal discourse with which growing inequality is associated, reduces social cohesion, which damages life expectancy and increases crime. It seems to many that Thatcherism, described by Gamble (1994) as 'the free market and the strong state', requires the strong state in order to lock up the extra criminals that are generated by operationalising the public choice – market part of the Thatcherite discourse.

The correlation between high crime rates, growing health inequalities and growing social inequality has been discussed by Wilkinson in some of his later work (1996). Kawachi and Kennedy (1998), who draw heavily on Wilkinson's work, summarise a series of points about income inequality, social cohesion and ill health.

First, they point out that studies in the United Kingdom and the United States suggest that the degree of 'income inequality predicts excess mortality within individual countries'. In one of these studies a 'Robin Hood' index of social inequality has been constructed to measure the degree of social inequality in a particular nation or region. Using this index the researchers (Kennedy et al. 1996) say that comparatively small rises in income inequality are associated with quite large increases in mortality.

Second, they show that the widening of the gap between rich and poor damages the social fabric leading to hostility between different sections of society and limited possibilities for social integration. Wilkinson has produced case-study comparisons. In the UK during World War II social solidarity led to dramatic improvements in life expectancy. However in the community of Roseto, Pennsylvania, widening income differentials and declining social solidarity were accompanied by increasing rates of coronary heart disease (Wilkinson 1996).

Third, there is evidence from political science research that the degree of social cohesion is correlated – through the strength of voluntary associations and citizen participation in local government – to the strength of democracy (Putnam 1993). If there is a creeping malaise in government – something that is certainly evidenced by very low voting turnouts among relatively poorer communities – then it may well be related to a creeping increase in social inequality.

Fourth, while there are big health differences between poor and rich areas, it seems likely that the sense of relative deprivation among poor people is a major cause of health inequality between people living in rich and poor areas. 'Not all studies have shown that poor people have worse health if they live in a poor area rather than a rich one...a poor person living in an affluent area may have a better environment but may also feel relatively poorer' (Kawachi and Kennedy 1997: 1039).

If this is the case then the effects of relative deprivation caused by geographical factors are increasing because there is increasing social segregation. I shall further discuss the issue of relative deprivation later.

Fifth, the growth of income inequality affects the quality of life of even those people usually counted, using the Galbraithian term, among the 'contented'. There is increased crime, declining motivations to learn and disciplinary problems in schools, declining faith in the political system and even, as a result of underskilling and poor education, suboptimal economic productivity and growth.

Kawachi and Kennedy's point about the increase in crime is supported by sociological research, in addition to the health-centred research on Russia mentioned earlier (Walberg *et al.* 1998) and findings produced by Wilkinson (1996). Witt *et al.* (1998), in a study on England and Wales, found that in the 1979–93 period the growth in earnings inequality and unemployment impact positively on five types of criminal activity examined. Fowles and Merva (1996), who studied links between income inequality in crime in the United States in the 1975–90 period found 'robust' links between wage inequality and the crimes of murder and assault.

Judge (1995) has questioned the basis of Wilkinson's findings concerning income and health inequality, but Wilkinson (1995) rejoined by saying that Judge had (erroneously) attacked only two out of five studies he had published and that many other studies conducted in both developed and developing countries also pointed in the same direction of strong correlations between income and health inequality. Wilkinson comments that behind the correlation between income and health lie a number of 'psychosocial factors such as sense of control, security, status, prestige, social distance and cohesion' (1995: 1285). Wilkinson's comments emphasise the point that while relative incomes are an indicator of social inequality, the causes of the inequalities in health are much more concerned, in developed countries at least, with non-material factors.

Since this particular skirmish, further studies have been produced showing strong correlations between income and health inequalities. One study (Kaplan *et al.* 1996), compared the annual mortality rates and the proportion of low income households in 50 states making up the United States. There is a clear trend for the states with higher than average income inequality to have higher mortality rates than those with relatively less income inequality. While the 25 states with the most income inequality had an average age adjusted annual mortality of around 887 per 100,000 people, those 25 states with the least income inequality had an annual age adjusted mortality of 821 per 100,000 people.[1] These figures are not inconsistent with a claim that a relatively modest effort to reduce social inequality could save an annual 75 per 100,000 people in the United States, or approaching 200,000 lives a year. The price of the pursuit of maximisation of material satisfaction through adopting more and more competitive practices and social division could be high.

Stress is said to be an important link between the existence of social inequality and poor health. We need to examine the phenomenon of stress.

Stress, health and work

When our brains sense a threat, chemical responses are triggered which help us deal with the threat in what is called a 'fight or flight' response. As soon as the threat is avoided the chemical response ends. But if the threat is chronic rather than acute, meaning that we cannot dispel the threat, a situation of psychological strain is induced. The chemical response lingers. This lingering response can lead to high levels of damaging chemicals being present in our bloodstream over long periods.

It is often said that this state of prolonged strain, or what I shall call 'stress', is a feature of modern life, where people are to a greater or lesser extent not in control. Although there are undoubtedly new stress situations in modern life, stress associated with subservience to others has always existed among humans and other animals and, as we shall see later, such stress can be associated with perceptions of inferior status.

Whatever the situation may be that causes stress, students of stress and the psychosocial responses that it induces suspect that those suffering from high stress over long periods are more likely to be susceptible to a wide variety of diseases. Certainly, the recent trend for relaxation therapy to be given to cancer sufferers in the belief that this improves their survival rates does, if such techniques are as useful as is believed, suggest that the reverse is true; namely, that stress leads to a higher risk of mortality from cancer. Stress counselling has been found to double the survival period for women breast cancer sufferers whose cancers have spread from the breast. According to Licinio *et al.* (1995), stress situations produce various hormones, including the corticotrophin-stimulating hormone (CRH), which by affecting immune cells reduce immunity to disease and, among other effects, produce genes that can turn normal cells into cancerous ones.

However, the most frequent studies of the relationship between stress and disease have been done relating to heart disease. This presents fewer methodological problems than cancer because it seems to be caused by much more short-term factors.[2] According to Karasek and Theorell, who have described studies suggesting strong links between stress at work and heart disease, high levels of cortisol and plasma catecholamine are associated with chronic strain and also hypertension, which is in turn associated with heart disease. These chemicals are released at times of stress (Karasek and Theorell 1990: 105-7). A link is thus established between stress, increased levels of chemicals which damage the circulatory system, and heart disease.

Karasek describes stress as being associated with high demands and a lack of control. Control, in the work situation, is described as 'decision latitude' which he says is 'interpreted as the worker's ability to control his or her own activities and skill usage, not to control others, although that is also a potentially important' (1990: 60). A senior architect has high decision latitude, but a worker on an assembly line, or a computer keyboard operator, has low decision latitude, low control and is therefore more vulnerable to stress. The popular vision of high-flying executives suffering from stress is untrue if realistic relative comparisons are to be made to the more junior members of the workforce, who have less control and status and suffer more stress. Unemployed people, despite their enforced idleness, have no jobs to have control over in the first place and they are also very low in status, an important factor that will be discussed later. Thus the involuntary unemployed will tend to have the highest stress levels of all.

Of course people have different skills and are suited to do different jobs. I would not be a very good architect or fireman, for example, although I make a tolerably good teacher and researcher. According to Karasek and Theorell (1990: 165) doing what is for a given person an ideal job as opposed to an average one will add seven years to that person's life expectancy. They are also likely to be much less susceptible to depression.

Although the studies seem to agree on the key importance of 'decision latitude' as an arbiter of stress, the contribution of high job demands to stress is controversial in the psycho-social literature. An article commenting on the results of a study on heart disease and stress in Czechoslovakia commented:

> It might be, for instance, that whereas more decision latitude is always better from the perspective of the individual, work demand

has an optimal range. If so it would function like alcohol consumption (or for that matter, like Goldilocks' porridge), in which a 'happy medium' is best. If true, this hypothesis would imply that those workplaces where high pace is a risk factor were, in relative terms, pushing their employers hard. On the other hand, work places where low pace is a risk factor were those where there was not enough work to go around.

(Bobak *et al.* 1998: 46)

This Czechoslovakian study found that, having adjusted the figures to take out the effects of age, the lowest quartile of workers, as defined by low decision latitude and job demand, were 25 per cent more likely to suffer from myocardial infarction (a form of heart disease) compared to the highest quartile of relatively unstressed workers. Consideration of risk factors such as smoking, lack of exercise and diet did not alter this conclusion (Bobak *et al.* 1998: 43).

The so-called 'Whitehall study' of civil servants which investigated the links between work-related stress and coronary heart disease (CHD) is the largest of its kind to have been conducted in the United Kingdom to date. It identified lack of control or low decision latitude as the key component in stress at work. The study, which involved around 7,400 civil servants monitored between 1985 and 1993, came to the stark conclusion that civil servants in the lowest grades were almost 50 per cent more likely to contract heart disease than those in the highest grades. Low-grade males were 50 per cent and low-grade females 47 per cent more susceptible to heart disease than their high-grade counterparts (Marmot *et al.* 1997). Some of this difference was accounted for the tendency of lower-grade workers to have higher conventional risk factors, mainly smoking (and also to a lesser extent, poor diet and little physical exercise), but more than half of the difference in susceptibility to heart disease between lower and higher grades was held to be attributable to job stress, this proportion being even higher in men than women (Marmot *et al.* 1997: 237). Extensive empirical evidence is presented by Karasek and Theorell (1990: 145–7) to suggest that there are links between heightened levels of smoking and boring and stressful jobs. There are thus good arguments to suggest that, directly or indirectly, most of the 50 per cent increased risk of heart disease among lower-grade workers observed in the Whitehall study is attributable to job stress.

The key importance of control to health may explain why people such as Winston Churchill and Lew Grade with unhealthy lifestyles but, on the other hand, high status and decision latitude managed to live to be in their nineties.

There are some further interpretations of the relationships between stress and health that could be reconciled with the results of the findings of the Whitehall study. It may be that in the case of the low ranking and clearly stressed civil servants that it may not merely be the low decision latitude of their jobs that gives them stress but also their low status. In fact it is difficult to separate out these factors. High-status jobs are invariably associated with high levels of decision latitude, and low-status jobs with limited discretion to use one's skills.

Another issue is that even in the case of relatively high-status jobs, it seems likely that employees can be put under stress if they have a poor relationship with their superiors. Constant denigration and perceived arbitrary restrictions on the ability of people to do their jobs are two possible high-stress situations. But it is difficult to study such effects since each situation is by nature individualised.

Bosses have been known to cast aside stress issues by suggesting that people who achieve high-control, high-status positions have a type of personality that helps them cope with stress, whereas those who occupy lower positions do so because of their inability to cope with pressure. Aside from the evidence that people of, say, a given educational standard, embrace a wide number of personality types and that educational attainment is a much more (although by no means perfect) accurate marker of status and control, Karasek and Theorell discuss considerable evidence to dispute the 'personality determines career' hypothesis. They cite various studies to support their position (1990: 114–16). Attempts to distract attention away from issues such as control over one's work patterns and status are symptomatic of a wider tendency to locate solutions for stress on the individual rather than organisational or social/political arrangements. All sorts of remedies are suggested to counter the effects of work-related stress. To be sure, some of the remedies suggested have a good basis in medical evidence. Physical exercise is correlated with stress relief, healthy diets and abstinence from smoking are all good in themselves. But authors such as Karasek and Theorell suggest that emphasis on the symptoms of stress is no substitute for tackling the root causes of stress.

Researchers such as Karasek have tended to concentrate on factors like decision latitude and job demand as factors that cause stress. However, they have also mentioned the issue of status as a factor. An important contribution to the debate about stress and its effects on society has been made by analysis of the question of status. I shall now turn to this.

Status and the competitive society

Oliver James (1997) has discussed the key role that status plays in maintaining well-being. Status affects all individuals and is deeply ingrained. James argues that the search for status has become dysfunctional under contemporary competitive capitalism. James describes how in primate communities seratonin levels in the bloodstream are related to status and how seratonin levels are associated with different patterns of behaviour, whether dominant or submissive. Submissive states which are associated with low seratonin levels involve states of depression and adopting submissive body language. This serves the function of communicating to the dominant individuals the acceptance of low status. Being in a state of submission involves a constraint on behaviour, a lack of control, and this induces a stress response.

Such patterns of behaviour have been well documented following experiments with ververt monkeys and also baboons. I feel uneasy about the practice of citing animal experiments. I do not see why that the research cannot be done with humans, even if it is rather more costly. A little research has been done, and is cited by James, although he relies heavily on research with ververt monkeys. I suspect that the reasons that it is not done with humans is that it would be far more expensive and that much more care would have to be taken to respect human sensitivities, which would again increase costs. However, in the absence of sufficient studies on humans I shall cite a couple of studies on baboons which indicate endocrine responses that are associated with dominance and submission.

Animals which are high up in the social hierarchy have been found to have much lower levels of chemicals associated with stress and damage to the cardiovascular system than submissive individuals (Sapolsky and Share 1994). Moreover, other studies suggest that strategies operated by submissive individuals to lower levels of damaging chemicals can involve initiating aggression. Aggression is identified as a displacement activity which gives temporary relief to stress induced by feeling of low status (Virgin and Sapolsky 1997).

James argues that since the 1950s people in the United Kingdom have assessed their own status much less on the basis of traditional roles but instead have embraced a much more apparently egalitarian way of assessing status measured largely through success in gaining material rewards. He cites the work of Veroff, who asked Americans identical questions about how they saw their roles in the 1950s and in the 1970s (1981). The replies illustrated a distinct shift in how people saw their role. Instead of their seeing status as being concerned with playing out a traditional role, they see it as pursuing individualised aspirations. The disadvantage with this development is that 'blame' for failure accrues to the individual.

The idea that society has become more individualised, with decisions on marriage, career, education being left more to the individual rather than decided according to the traditional family and community structures, forms an important part of 'risk society' analysis put together by Ulrich Beck, who comments directly on the increased role of competition:

> Competition rests upon the interchangeability of qualifications and therefore compels people to advertise the individuality and uniqueness of their work and of their accomplishments. The growing pressure of competition leads to an individualisation among equals, i.e. precisely in areas of conduct which are characterised by a shared backgrounds (similar education, similar experience, similar knowledge). Especially where such a shared background still exists, community is dissolved in the acid bath of competition. In this sense, competition undermines the equality of equals, without, however, eliminating it. It causes the isolation of individuals within homogeneous social groups.
>
> (Beck 1998: 94)

The individualisation of blame leads to widespread stress and depression. This has been a general trend in western society, but James detects that the United Kingdom has become more 'Americanised' in the sense that it has adopted its own implicit version of the US slogan 'anyone can be President'. The sense of alienation and relative deprivation is worst when people come from backgrounds that are relatively unequal in terms of material status, but are assessed by the same materialist and allegedly egalitarian standards as others from backgrounds where relative success in educational and material terms is more the norm. People's sense of well-being has stopped improving. People are encouraged to set themselves lofty goals, and instead of comparing themselves with those who are less fortunate they tend to compare themselves upwards. Instead of thinking of reasons why they are better than other people, they think about how they are inferior to others, even when they may be a member of a relatively highly placed peer groups. Their reference group, the group whose goals they aspire, is often out of their reach. This produces depression and unhappiness.

Of course, people at the bottom are particularly badly affected, in status terms. They see themselves as much more deprived by contrasting what seems to be successful consumer society with their own existence. No matter how rich they may seem, in absolute terms, compared with their forebears, they still suffer from a sense of deprivation relative to others. Relative deprivation is a concept that is central to the logic contained in this work.

Government surveys of whether people feel happy suggest that people are still about as happy as they were 20 or 30 years ago, something that confirms the notion that absolute gains in material standards of living over this period have not changed the levels of general happiness. Two economists who 'monetarised' happiness – Blanchflower and Oswald (1999) – accept these general findings. However, this does not stop them from declaring that they had discovered a formula for happiness which relies on possession of money as the central variable. 'Money does buy happiness', their report declared, a finding that was widely reported in the British press. They leave no clear argument to dispel the apparent glaring contradiction between their claim that there is a causal link between happiness and income and the fact that happiness has remained flat or even declining in the United Kingdom and the United States over recent decades. Their alleged (and empirically dubious) link between income levels and happiness provides justification for dominant neo-liberal discourse that concentration on the maximisation of financial returns to the individual will assure maximum happiness when the degree of happiness of individuals is added together.

The point which the authors appear to have missed is that it is not the money itself which influences individual levels of happiness, but the status and decision latitude relative to others. Money is associated with happiness and unhappiness on a *relative*, not *absolute* basis, so that more money does not make western societies, as a whole, happier as per capita GDP increases. This suggests that the solution to the problem of maximising happiness, or the balance of pleasure over pain (to put the problem in Benthamite utilitarian language), is not therefore to induce people to compete more vigorously with each other for more money. This merely makes some people, 'the winners', happier at 'the losers'' expense. Rather, the solution could be concerned with a reduction in the importance of money as an indicator of status, and an attempt to promote the notion that status can be more widely achieved through co-operation and community membership.

There are, of course, faults in the utilitarian theory about maximising pleasure for as many people as possible, apart from the widely discussed notion that what makes the majority happy does not necessarily appeal to a minority. A further problem is that there is no attempt to investigate the origins of the preferences upon which judgements of pleasure and pain are based. If, for example, contemporary social norms favour pleasure being derived from fast, gas-guzzling cars and going around having fights to prove masculinity, then utilitarian theorists (and rational choice theorists whose logic on this and other points is shared) are unable to do anything more than pronounce that 'good' is concerned with promoting this set of conventional norms. In this case it is conventional norms – what is assumed to be self-evidently the individual's rational choice – that is the problem. Similarly, the notion that rationality is concerned with competing on an individual basis rather than co-operating is now the conventional norm in the Anglo-Saxon world, and that is the problem. The culture of lauding the notion of aiming to win rather than co-operate is a strategy that ignores the damaging effect of the stigma attached to being seen as 'a loser'. Everyone likes to be a winner – there is no problem here – but the problem is that not everybody can be a winner under the super-competitive system.

Competition and consumption

Baumann (1998) has discussed how western society has moved from conceptualising individual identity as based on the work ethic – which involves co-operation and co-ordination among producers – to an ethic of consumption:

> The activity of consumption is a natural enemy of all co-ordination and integration. ...Consumers are alone even when they act together. ...Freedom to choose sets the stratification ladder of consumer society and so also the frame in which its members, the consumers, inscribe their life aspirations – a frame that defines the direction of efforts towards self-improvement and encloses the image of 'a good life'. The more freedom of choice one has, and above all the more choice one freely exercises, the higher up one is placed in the social hierarchy, the more public deference and self-esteem one can count on. ...The prime significance of wealth and income is the stretching of the range of consumer choice.
> (Baumann 1998: 30–1)

Schor (1998) adopts a similar approach when she examines what she calls 'the new consumerism' that has taken hold in the United States since the 1980s. She points to the trend away from people comparing themselves with, and thus basing their consumer decisions upon, their own peer groups and towards people comparing themselves to super-rich reference groups:

> The result is that millions of us have become participants in a national culture of upscale spending. ...When twenty somethings can't afford much more than a utilitarian studio but think they should have a New York apartment to match the ones they see on *Friends*, they are setting unattainable consumption goals for themselves, with dissatisfaction as a predictable result. When the children of affluent suburban and impoverished inner-city households both want the same Tommy Hilfinger logo emblazoned on their chests and the top-of-the-line Swoosh on their feet, it's a potential disaster.

She recounts a discussion, on a talk show programme about what dodgy methods one can adopt in order to appear to be rich, and adds, 'Apparently the upscale life is now so worth living that deception, cheating, and theft are a small price to pay for it' (Schor 1998: 4–5). This 'upscaling' spending tendency is also related to what Schor calls the 'Diderot' effect. Just as the eighteenth-century Diderot described how, on buying an elegant gown, he felt necessary to 'upscale' his surroundings to match his new, expensive piece of apparel. So, today, consumer pressures push people towards a never-ending cycle of spending to buy objects and settings that fall in line with an initial expensive purchase. A new home is not complete without new expensive furnishings; a new, pricey skirt is not complete without an equally high-class jacket; and so on. What is marketed as a means of individual expression is also pressure for social conformity on an upwardly moving escalator of increasing personal spending (Schor 1998: 145). While all sections feel the pressure to keep up consumer appearances that results from this trend, the stress is severest among those who have least ability to compete in this pattern of conspicuous spending – conspicuous in that the purpose of the spending, for example, through possession of Rolex watches, is solely to show one's superior place in the new competitive hierarchy of consumption.

The consequences of too much competition

James, looking at the United Kingdom, describes much the same trends of increasing pressure to display success in material terms. He also discusses the psychological consequences in terms of how feelings of relative deprivation have increased, material inequality has widened and society has become more violent. James comments that the increase in inequality 'has greatly increased the proportion of boys raised in low income homes which are most likely to provide violence-inducing childcare' (1997: 27).

The theory of relative deprivation and its relation to social justice was discussed in the 1960s by Runciman (1966). Ironically, this was a few years before the neo-liberal discourse began to intensify relative deprivation. The growth of relative deprivation confirms the identity of a whole class of people as people who are at the bottom of the heap. The educational system, the legal and judicial system, mainstream commerce and industry are not for them. These make up the system which has no place for the underclass and which often acts positively to oppress them. It is little wonder that poor areas are increasingly associated with underperformance in education, given that children of families in the underclass see the educational system as not helping them. The children derive their identities from their parents and their peers, and this culture suggests that they will not win and they will derive little from the educational system since others who share their identity did not 'succeed'. It is also little wonder that poor people have inferior diets, smoke more and generally enjoy a less healthy lifestyle than their middle-class counterparts. They simply have little faith in, confidence in, and desire to participate in a system – even the part of it dealing with advice on health and lifestyle matters – that is not seen as acting in their favour.

Many members of the underclass find that their health is attacked on two fronts. First, there is the stress prompted by lack of status and lack of decision latitude. These conditions affect the unemployed particularly strongly, and the stress produces physiologically damaging consequences, which, as explained before, leads to immune deficiency and damage to the circulatory system. Second are the consequences of the unhealthy lifestyles that result at least partly from the alienation of members of the underclass from the 'system', and their lack of faith in the notion that the system can help them improve their lifestyles.

However, the problem of relative deprivation does not just affect the poor and the most oppressed workers. It reaches right through society. However well off in material terms people may be, they are susceptible to relative deprivation. Expectations have dramatically increased, producing a society of 'wannabes', the majority of whom are destined to find that their expectations have not been met.

In Chapter 4 I discussed, with the help of Galbraith, how a perception by the growing numbers of middle-class people that they would be better off paying lower taxes has helped the neo-liberal discourse appear resonant. An additional factor that aids this resonance is that the neo-liberal discourse, which emphasises competition, appears to coincide with the breakdown in traditional roles and the assessment of status through individualised competition. However, the costs of such developments are accepted as an inevitable part of what is regarded as progress.

Even when people are in the more advantaged middle classes, the drive for competition encourages depression and stress associated with a lack of status when people do not achieve their full ambitions. Even apparently highly successful people can feel failures. James uses the Royal Family as an example and looks at the cases of Princess Diana and Sarah Ferguson, both of who seemed to assess themselves on criteria other than their success in marrying into the pinnacle of the nation's traditional elite.

The new discourse of stress at work

Not only do the apparently successful middle classes suffer from feelings of low status when they do not achieve their elevated ambitions, but they can also be stressed by bad relations with bosses who themselves may suffer from feelings of relative deprivation from failure to achieve exacting performance targets. Indeed, a 'stress discourse' has become an increasing facet of middle-class fears. For example, the publication of a study by the Institute of Personnel Development led to a storyline: 'Overworked and stressed managers are replacing bosses with personality defects as the main culprits in a spreading epidemic of workplace bullying' (Milne 1998).

This particular report estimated that one in eight workers had been bullied in the previous eight years and that managerial staff were just as likely to be bullied as others. 'Bullying typically consists of unfair and excess criticism, public insults, repeatedly changing or setting unrealistic work targets, undervaluing of work efforts, shouting and abusive behaviour.' We have seen the beginning of a trend towards litigation by alleged victims of stress who have, in some well-publicised instances, received significant amounts of compensation.

The position of some members of the workforce is only a little better than that of the underclass in that their working conditions are highly stressful. Any notion that new technology will abolish repetitive work and the dull regimentation of line management are questioned by the experience of call-centre (automated distribution systems or ACDs) workers. The number of call-centre workers who deal in all sorts of services ranging from banking to airline ticket sales has been rapidly increasing in recent years. The new technology associated with call centres makes it all the more easy to use electronic monitoring of performance. Electronic measurement of performance is something that has been identified as an especially stressful technique which nevertheless is almost the ideal way of implementing the neo-liberal ideal of treating the employee as a unit of production whose performance is perfectly transparent. It is also the ultimate stressor, in that the employee has no control over the means and timing of the surveillance and must remain in a state of permanent fearfulness that an Orwellian style Big Brother will pick up their every mistake, foible and inconsistency.

Call-centre workers face most of the classic promoters of stress. They have little control over the pace of work, they are subject to arbitrary monitoring, and their very individuality is threatened as they are transformed into routinised semi-automatons with little personal space or freedom. Just as animal rights campaigners complain about the stresses facing battery hens, so one might complain about the position of call-centre workers whose conditions are paralleled in many senses by battery hens. Staff turnover rates tend to be as high in the traditional production-line factory. One researcher comments:

> What we have found is that there is a hugely significant relationship between levels of perceived control over the pace of work and people's job satisfaction and stress. As you would expect, if people have low control over their jobs, they experience high levels of stress and low levels of job satisfaction, so there is a big problem in call centres generally.
>
> (Arkin 1997: 24)

Of course many others – probably a majority of workers – suffer such problems to a greater or lesser degree. A key problem is that some elements of what is described as bullying – such as setting unrealistic targets and undervaluing work efforts – is increasingly institutionalised in work through performance appraisal, performance targets and performance-related pay. The stated aim of the drive towards competition may be to increase productivity, but an effect is to enhance the possibilities for relative deprivation, loss of control and status, increased stress, and of course the damaging impacts on health that this involves.

Burchell et al. (1999), in a report for the Rowntree Trust, have highlighted these trends. Performance appraisal systems, 'a prevalent feature of modern organisations...tend to increase stress and decrease well being. In particular, being subject to individual performance targets, especially when these targets are unachievable, seems to exert a negative impact on employees' general health and is associated with negative feelings about work' (1999: 45). Job insecurity produces a decline in well-being, and 'long term insecurity seems to build up and cause a continuing decline in well-being. This suggests that, if the UK economy continues to generate high levels of job insecurity, the stress related problems may continue to increase' (1999: 46).

The Rowntree Trust Report commented that there was widespread mistrust of management: 'Largely *unconditional* demands are being made of workers by their employers whilst the commitment to job security, and the other promises by employers to their employees are *conditional*' on not only (unavoidable) economic uncertainty but also (avoidable) organisational restructuring. 'The unconditionality of demands and the conditionality of promises mean that workers are required to be totally committed to organisational objectives, yet readily disposable. And whilst the former strives to collectivise effort, the latter de-collectivises and individualises risk' (1999: 61). This analysis is redolent of Beck's description of the move towards a 'risk society'. The Report also touches on ecological ways of thinking, and fits in with the general approach followed in this work, when it comments:

> Like environmental scientists point out that many organisations fail to take account of the damage they do to the *natural environment*, what we as social scientists must stress is the extent to which job insecurity and job intensification damage the *social environment*. We have seen that many of the organisations in our survey claim to have profited from the downsizing and restructuring of their workforce. But the costs of these developments, in terms of greater insecurity and reduced physical and psychological well-being, do not appear on their balance sheets.
> (Burchell et al. 1999: 62–3)

There is a criticism to make of this particular report. It is in terms of its implicit assumption that the techniques of job intensification and job insecurity are of benefit to the individual firm, despite the externalised costs to society. As we shall see in Chapter 7, when we come to look at some research on solutions to job-related stress, even the alleged internal economic benefits said to be associated with stress-inducing neo-liberal techniques are in doubt.

It would be wrong to ignore the doubts about the new stress discourse. Criticisms can be put into three categories. First is the claim that while stress may have emerged as a concern in the recent past, in fact nothing has changed, thus suggesting that the notion of stress is a mere social or discursive construction without any real substance. I would agree that stress is a discursive construction, but so – as will be made clear by the discussion in Chapter 8 – are many if not most things. Nevertheless, as I discuss in Chapter 8, we can talk about a reality concerned with the health and survival of our particular species, a physiological reality that is common to all humans. The evidence, some of which has been cited in this chapter, suggests that stress does have a 'real' basis in that it has physiological consequences which can shorten lives. Neither does the fact that it is a recently discovered notion reduce its importance. As is discussed elsewhere in this work, concerns about industrial pollution have only been given a really high priority since World War II, even though such problems have existed in various forms down through human history. However, the relatively recent onset of concern for the environment as a general, global problem does not devalue the importance of dealing with pollution.

Second is the criticism of the 'stress management industry', which is said to be producing mistrust and the loss of 'confidence in people's ability to cope with the normal challenges associated with the world of work' (Furedi 1999). I would certainly agree that the emphasis placed by many companies on stress management schemes and the encouragement of stress counselling represents an effort to deal with the symptoms of the problem rather than the cause. The point is that the work environment has to be changed, along with people's attitude to work. The neo-liberal discourse, which sanctions excessive competition, needs to be abandoned. These themes are discussed in much greater detail in Chapter 7. Stress management may be superfluous, but we still need to tackle the structural factors that produce stress.

A third type of criticism, coming from a Marxist perspective, may be that in so far as stress is a problem, it is a consequence of the class oppression of workers that is endemic under capitalism, and that it is the capitalist mode of production that needs to be changed. To the extent that it is the most oppressed sections of society that are the most stressed and that it is these who are least likely to have their stress problems addressed, I would agree with some of this criticism. However, stress can exist independently of the economic organisation of society.

Crucially, a different mode of production based on a socialised economy is not in itself guaranteed to resolve the problem, especially if it promotes the pursuit of economic growth as a higher objective than other indicators of quality of life.

We can see from the evidence presented in this chapter that psycho-social stress – that is, stress that is generated as a consequence of the way society is organised – can be a significant external cost of material production and consumption. Just as some forms of economic growth may be more damaging to the environment than others, through the consumption of more resources and the production of more pollution, so some forms of economic growth (those that produce social inequality, in particular through excessive competition), are likely to generate significant external costs for society that are a consequence of increased psycho-social stress.

In the next chapter I shall examine the case of performance-related pay, in order to illustrate the impacts of the trend towards marketisation which involves, at its heart, turning people into more and more competitive agents.

6
The Politics of Performance

Foucault discussed the nature of political control through the concept of 'governmentality' (1991). This concept of governmentality was said to be operationalised through a set of institutions and professions, a set of techniques and a set of knowledges or discourses that underpinned these techniques. Neo-liberalism is such a discourse that has, associated with it, a number of techniques that are put into practice by a varied number of agencies. I have discussed some of the knowledge constituted by the neo-liberal discourse in Chapter 4, and now I want to look at a couple of the institutions and techniques associated with neo-liberalism. A key difference between neo-liberalism and its nineteenth-century *laissez-faire* antecedent is the role, under neo-liberalism, accorded to the government to ensure that the right set of institutions and techniques are adopted and applied. These institutions and techniques, some of which I shall now examine, are aimed at regulating and promoting competition. First, I shall investigate the impact of performance-related pay, a technique associated with – ostensibly at least – the aim of inducing employees to work harder and more efficiently. Then I shall look at schools in the United Kingdom and discuss aspects of the work of institutions like OFSTED and techniques such as school league tables.

Performance-related pay (PRP), which involves (according to its theory) paying people according to what they have produced, spread inexorably throughout the US and UK economies in the 1980s and 1990s. By 1998, in the United Kingdom, around two-thirds of private sector companies had some form of performance-related pay (Rafferty and Barnard 1998). The practice had penetrated the public sector as part of the Thatcherite effort to streamline the Civil Service, a policy that was specifically endorsed by Labour Chancellor Gordon Brown soon after gaining office in 1997.

Performance pay flows from the logic of the neo-liberal discourse in that the promotion of competition through the creation of intra-organisation markets for output should, according to theory, increase productivity by setting up 'tournaments' among employees. But does this produce the sort of stress-related external costs of production discussed in Chapter 5?

There is a growing academic literature on PRP and, although there are a number of studies on the theory of how PRP should work, the literature does tend to suggest that the only cases where there are clear gains in productivity resulting from PRP are activities, such as direct selling, where performance can be measured with a high degree of accuracy – or at least there is agreement among the interested parties. This situation, however, is not typical of even the private sector, and moreover is akin to piecework, which has been identified as being a particularly stressful way of paying people.

A study of the impact on productivity changes following the introduction of a piecework scheme in a large US autoglass company in the 1994–95 period revealed that this type of performance-related pay scheme did indeed improve productivity between 20 and 36 per cent. About half of the increase in productivity was passed on to workers in higher wages (Lazear 1996). On the other hand, studies suggest that there are large 'external costs' associated with stress involved in such schemes. Levi (1972) found that raised blood catecholamine levels (associated with the development of heart disease) are associated with shifts from payment by salary to piecework. Karasek and Theorell (1990: 62) comment:

> [A] weekly salary is actually a source of security for a worker who might otherwise be exposed to all manner of management-initiated output pressures, such as competition among co-workers and new production goals. Piecework's 'freedom' to work harder is rather like the freedom of the poor to sleep under bridges: it brings with it a set of new demands for adjusting to the environment that may more than offset the salubrious effects of control.

Outside of cases where increases in output can be clearly measured according to widely agreed criteria, the literature produces verdicts on PRPs effects on productivity that are at best mixed, and often negative (Zenger 1992; Hede 1991; Lowery 1995; Walsh 1993; Gregg 1993; Rosen 1990; Marsden and French 1998). The conclusion that PRP demotivates significant numbers of employees seems to be near universal.

Walsh (1993), who investigated PRP schemes at a number of large companies, concluded that such schemes could actually raise the costs of managing the employer–employee relationship and that PRP could act to demotivate staff. She observed that in conditions of boom such as the late 1980s many schemes actually led to increasing staff costs and damages to team working. However, PRP proved more useful (to bosses) in recessions as it enabled employers not to pay cost-of-living increases or salary increments to employees whose performance was judged to be below par (1993: 422).

Interestingly, this contrasts with assessments of schemes to link the pay of company directors with performance as measured by profits. Surveys in both the United States and the United Kingdom report only a weak connection between directors' pay and company performance (Rosen 1990; Gregg et al. 1993). Gregg et al. found that directors' compensation was 'insignificantly related' to corporate performance and that directors' pay increased at rates of around 20 per cent per year during the period covered (1983–91) – far above those accorded to other employees. In times of recession the connection between directors' pay and profitability entirely disappeared, or even became negatively related. It should be noted that during recessions while PRP schemes were continuing to award large increases to company directors, Walsh (1993) reported that PRP schemes were being used to deny many employees any increases at all.

Gregg et al. observed that the pay of company directors was driven 'more by size than by performance', meaning that 'directors have a clear incentive to pursue merger and acquisition activity regardless of any benefit to shareholders, workers or the economy as a whole' (1993: 2).

Although I would agree with authors such as Smith and Morton (1993) when they point out that techniques such as PRP have the effect of marginalising trade unions by reducing their influence over wage bargaining, the evidence produced by Gregg et al. does not support Smith and Morton's neo-Marxist interpretation that the purpose of the techniques is to 'permit an intensification of the rate of appropriation of surplus value over surplus labour' (Gregg et al. 1993). The motivation of directors seems to be growth in turnover rather than increases in profits. An alternative interpretation would be that the employers are utilising the neo-liberal discourse to legitimise strategies which increase their own control and also to increase greatly their own rewards in the context of a widening of income inequalities.

UK surveys of the impact of PRP schemes in large utilities and the public sector paint a bleak picture of the effects on employee morale. Large majorities are reported as being demotivated. The Society of Telephone Engineers, in a survey of staff at British Telecom (2,600 replies), found that only 6 per cent of staff felt that the PRP scheme had raised their motivation and that 84 per cent felt that the pay system was unfair (STE 1996: 33). There was evidence of stress-related symptoms among those who worked long hours, with, for example, two-thirds of respondents working over 55 hours having difficulty sleeping compared with 37 per cent who worked 41–42 hours (1996: 48).

Marsden and French (1998) have undoubtedly produced (at the time of writing) the most rigorous survey of PRP experience in the public sector in the United Kingdom, and its findings were damning. Extensive and highly detailed surveys were conducted at the Inland Revenue, the Employment Service and sections of the health service and schools where PRP schemes were in operation. According to the results, most staff believed that PRP did not raise their motivation, that it is divisive, undermines morale, causes jealousies and inhibits workplace co-operation. Only a minority of line managers believed that PRP had improved productivity and many staff believed that the schemes were used by line managers to reward their own favourites.

It is apparent from the literature on PRP that it reinforces, and in some cases creates, the winners-and-losers status system described in the last chapter, where people are forced to compare themselves with others on the basis of an allegedly egalitarian system of assessment that creates stress and depression for those who do not come out on top.

Yet, despite the apparent ineffectiveness in promoting productivity gains and despite the widespread tendency of PRP to demotivate employees, the Marsden Report says: 'One of the most striking responses across our case studies was the degree of support for the principle of linking pay to performance' (1998: 9). Only in schools was there a majority against it.

The contrast between the implicit criticisms of the practice of PRP and the 'in principle' support for such a system is given in Table 6.1, where I present a selection of responses concerning PRP drawn from the survey of Inland Revenue staff conducted by Marsden and French.

The general acceptance of the principle of PRP has undercut any idea unions might have had to oppose PRP schemes (teachers' unions aside). The point is that the dominance of the neo-liberal discourse legitimises PRP schemes since they operationalise the discourse's sanctioning of greater competition. PRP is associated in people's minds with progress and modernisation. But the fact that people currently accept that their self-interest should be shaped by the neo-liberal discourse does not mean that there is no potentially plausible alternative interpretation

Table 6.1 Attitudes to performance pay among UK Inland Revenue staff

	Disagree (%)	Agree (%)
Questions to staff		
The principle of relating pay to performance is a good one	37	58
Pay should reflect the demands of the job and not the performance of the individual job holders	29	58
People should be paid according to nationally negotiated pay scales	13	76
Performance management makes staff less willing to assist colleagues experiencing work difficulties	30	63
Performance management causes jealousies between staff	9	86
Performance management has helped to undermine staff morale	10	81
Performance management discourages team working	23	67
Performance management has caused greater stress in my job	26	64
Questions to line management		
Performance management has led to an increase in the *quantity* of the work many of the staff do	51	42
Performance management has led to an increase in the *quality* of the work of many of the staff	72	16

Source: Marsden and French (1998); selection taken from tables 2.5–2.19

of self-interest based on a different, perhaps less materialist and certainly less competitive, discourse. The practical experience of PRP schemes attests to that.

The ineffectiveness of trade union efforts to oppose PRP and, similarly, the relentless drive towards more and more competition and marketisation, is related to the materialist left's general inability to articulate a critique of the external costs of production. In this case the external costs are the stress and social division that results from the competitive techniques operationalised in the name of the neo-liberal discourse.

One area that has been subjected to increasing competition in recent years is the schools sector. It provides a good example of the pitfalls of the marketisation process.

Education, competition and performance

The drive towards marketisation of schools seems to be much faster in the United Kingdom compared to the United States. For example, whereas the United Kingdom now has a system of performance-related pay for teachers in all public sector schools, PRP is still relatively rare in American schools, although, as this book goes to press, US teaching organisations are watching to see what happens following the inception of a PRP scheme in schools in Denver, Colorado. In the United Kingdom there has been a tremendous growth in institutionalised competition following the 1986 and especially the 1988 Education Act. The Conservative Government attempted to establish a market in schooling by allowing parents, on paper at least, the freedom to choose the school which their children should attend and also by attempting to give information to the parents on the relative academic success of different schools. League tables of GCSE and GCE results were published from 1991 onwards. Comparison of schools could be achieved only by trying to ensure that the same product was being manufactured, hence an added rationale for the national curriculum and the introduction of an apparently rigorous regime of national testing at ages 7, 11 and 14.

Alongside the new array of measures to standardise and to compare children and schools, there has been a drive to improve the performance of teachers. The concept of the 'failing' teacher has been institutionalised by the Office for Standards in Education (OFSTED). OFSTED, established in the early 1990s, announced that there were, as a conservative estimate around 15,000 failing teachers in the United Kingdom (Macleod 1996), and that a third of schools 'are not as good as they should be' (DFEE 1997: 10). OFSTED was an important step in extending the system of institutionalised performance comparisons and competition to the teachers themselves. Four explanations for this new system can be put forward. First, it is in keeping with the neo-liberal discourse's suspicions that, at least left to themselves, publicly funded activities inevitably end up being inefficient and self-serving. OFSTED was thus needed to prevent this parasitic activity.

Second, putting more emphasis on the role of the teacher and the pupil to perform better switches attention away from the role of society, through peer groups and social background, in influencing, and as we shall see later, mostly determining educational achievement. The neo-liberal discourse assumes that the prime motivation of humans is for individual material gain and that in education, as in other situations, the individual will respond effectively to the correct material incentives in a competitive context. Government policy explicitly loads responsibility for performance onto teachers and pupils: as the *White Paper Learning to Compete* (DFEE 1996) put it, 'the action to achieve the success we need rests principally with young people themselves, and with those who teach, train, employ and advise them' (1996: 49).

A third explanation for the shift of responsibility is that, because educational attainment has become such a measurement of status, this increases the desire of parents to evade personal responsibility for perceived (and often purely relative) by loading 'failure' onto someone else.

A fourth possible explanation is the shift in power to the richer sections of the middle classes who wanted the freedom to send their children to the schools of their choosing, a feat they could achieve by leap-frogging over other parents by being able to buy themselves (through property purchases) into the areas close to the schools with the most middle-class intakes.

The Labour Government not only accepted the marketisation of primary and secondary education pursued by the Conservatives, but even announced a desire to extend it, justifying this with the observation that 'We face new challenges at home and from international competitors, such as the Pacific Rim countries' (DFEE 1997: 11). The Labour Government accepted the neo-liberal discourse, although the texts of Government policy documents reveal some conflicts with the Labour Party's traditional (and mostly abandoned) commitments to equality. Its 1997 *White Paper* did recognise the importance of parental involvement in children's education (DFEE 1997: 12), and in 1998 the *Green Paper Teachers – Meeting the Challenge of Change* (DFEE 1998) commented that 'We will recognise and provide support for schools facing economic and social disadvantage, but this cannot be allowed to be an excuse for under-performance.' Nevertheless the policy still insists that social disadvantage can and must be overcome by the efforts of schools, and that if it is not, the failure will be publicly exposed and 'failing' schools closed (DFEE 1997: 12).

The Labour Government have justified the extension of competition into schools on the basis of the same 'modernisation' discourse that has justified the introduction of performance pay elsewhere. Teaching, of course, which has involved a caring attitude not conducive to regarding children to be looked on as units of resource to be processed, and which has involved inculcating children with the benefits of co-operation with, and caring for, fellow citizens, is not the easiest part of the economy to assimilate into the super-competitive system. The

Government implies that those who dislike the new regime of performance pay which the Government announced in late 1998 are out of step with what is called the 'the imperative of modernisation' (DFEE 1998: 11). As the *Green Paper* stated:

> Teachers too often seem to be afraid of change and therefore to resist it. Teachers have too often felt isolated. many seem to believe that they are unique victims of the process of constant change, although the reality is that in many other sectors change has been more revolutionary and had greater impact on pay, conditions and styles of work.
>
> (DFEE 1998: 16)

It is certainly the case that the marketisation of education is in step with the proliferation of performance-related pay and the increase in the institutionalisation of competitive behaviour in the rest of the so-called productive economy, but the assumption that these changes are necessarily beneficial is highly contestable.

Does increasing competition in education work?

The new policies which aim to improve performance of pupils by increasing the amount of competition in the educational system can be criticised on many grounds. A fundamental criticism is that it has institutionalised an already visible trend for education to become a battleground for competitive individualism, a battleground that determines the status of children, and by association, their parents, on the singular basis of academic attainment. Academic success, of course, is linked to the drive for material success. But like the competitive materialist individualism that has been made ever more the measure of social status, this competition creates losers who become alienated from the system at an early (and it seems increasingly earlier) age. James comments: 'There is a large body of evidence that social comparison and competition at school cause a lowering of self-confidence, optimism and self-esteem in most children starting from age seven onwards' (James 1997: 117).

James cites the research of James Ruble on how this loss of self-esteem among children leads them often to operate strategies of disengagement from the system, and of seeking alternative forms of status in terms of 'anti-school' cliques and gangs. It follows from this that the more emphasis that is paid on academic performance as measure of status, then the worse will be the sense of failure among the 'losers' in the competitive race. The resulting alienation can provide a powerful part (in addition to the trend towards divorce and broken homes) of the explanation for the widely perceived, and in terms of rising numbers of school expulsions, empirically observed, trend towards bad behaviour and disruption of lessons in recent years.

There is much empirical evidence suggesting that the goal of raising standards of educational achievement is not being achieved, and that the competitive theory upon which this drive is based is deeply flawed.

A study of schools inspected by OFSTED, commissioned by the Liberal Democrats, suggested that, if anything, OFSTED has damaged the absolute performance (in terms of improvements to GCSE results) of those schools that were inspected compared to those that escaped, for the time being, this privilege (Carvel 1998). Schools are forced to devote a great deal of effort to fend off the threat of a bad OFSTED report. There is considerable evidence of great injustices done as a result of this system, with schools whose results (of work done just before the inspections) turned out to be good by comparison being publicly pilloried, and relatively 'underperforming' schools being lauded (Aitkenhead 1998).

GCSE grades have risen consistently since the mid-1980s. One can debate the extent to which this represents a genuine improvement in learning or the extent to which this represents how syllabuses have been shortened and simplified and pass-grade boundaries reduced. What is clear is that the differences in the results between the 'successful' and the 'failing' schools and pupils widened consistently during the 1990s, up to at least the time of writing. As a *Guardian* leading article commented: 'a rising number failing to get any GCSE passes and a widening of the gap between the top and bottom are problems that are not new but remain the biggest challenge facing ministers' (*Guardian leader* 1998).

Despite the commitments of successive governments to improve standards of literacy and numeracy, such standards have remained largely the same for the past 50 years (Brooks 1998). According to the National Literacy Association, standards of illiteracy have remained virtually constant, at 16 per cent, since 1950. Given the research on the factors that influence degrees of learning among children, it is not surprising that the tactics of increasing competition should not affect this statistic.

Given this evidence, it is reasonable to conclude that the effect of the Government's ruthless effort to stamp out failure is not only ineffective, but also counter-productive and socially damaging. The effect of the increased competition, and the various indicators and devices that promote it, is to emphasise the difference between success and failure, to reduce self-esteem of the 'failures', to increase stress generally and to increase social division.

The problem of 'low esteem' is recognised in Government policy documents, and small projects such as the 'Make a Difference' scheme have been launched 'to improve young people's self esteem and motivation though involvement in voluntary activity' (DFEE 1996: 33). However, because success in such schemes is not valued as an important indicator of success by society – as evidenced, for example, by the lack of 'league tables' for 'success' in voluntary work – so status is seen as being in effect solely determined by the degree of academic success. This is deemed likely to contribute to the paramount goal of economic growth, while voluntary work does not. It is not surprising that, in general, those who benefit most from the gains in such economic growth have children who are successful at school.

Mainstream academic research carried out using data on literacy and numeracy standards suggests that, as Robertson and Symons put it, 'to a first approximation at least, the academic attainment of children in schools is completely explained by the education, income and social class of their parents; and by the levels of these factors in the child's peer groups' (1996: 1). Robertson and Symons' own work suggests that peer group influence, based on the values of the peer group and attitudes to education, exerts a significant influence independent of economic class, so that (for instance) a working-class child will do better in a middle-class school than in a working-class school. However, the source of the attitudes that generate the peer group values is, in the individual case, still essentially based on social and economic background.

This class-based theory of educational attainment questions the relevance of both new right and traditional socialist ideas on improving standards. The new right (whose ideas have been accepted by the Labour Government) emphasise improving teaching methods through competition, both external and internal to the school. But as Brynner and Steedman put it, 'the effects of social and economic disadvantage are so dominant in impeding basic skills development that they overwhelm the "fine tuning" represented by school organisation and curriculum' (1995: 67). Robertson and Symons comment that 'School quality as measured by conventional inputs such as class size, teacher experience and general school expenditure are typically found to have only minor effects' (1996: 1). It seems that the left-wing 'answer' of improving standards simply by spending more money is not going to be more than marginally effective either.

To make such claims is not to say that individuals from poor social backgrounds cannot ever achieve educational success or that some teachers are not better than others. But the motivation for educational success must exist inside the person, and those from a poor background who do succeed gain their motivation despite the often poor faith in the system held by their parents and peers. People tend to derive their sense of identity and role from their family and peer group, and if the knowledge so derived is that the system does not deliver results for them, they will not be highly motivated to collaborate with a system that is being perceived to act against them.

It seems likely that the moral regulation described by Durkheim that provides society with organic solidarity is in decline. Durkheim said that 'It is by respecting the school rules that each child learns to respect rules in general, that he develops the habit of self-control and restraint simply because he should control and restrain himself' (Durkheim 1961: 149, cited by Ball 1981: 314). But what Durkheim said may also work in the other direction. When people respect society less, especially when its leaders, such as Mrs Thatcher, claim that it does not even exist, then the children are likely to respect the school rules less.

If the preceding line of argument is correct then the new Government emphasis on 'zero tolerance on underperformance' and threats to close allegedly failing schools (DFEE 1998: 12) can only damage society. The sin of failure, already very implicit in the educational system, is now reified and excoriated. In the ever more competitive system the existence of failure is made more and more explicit and the low status of some who must always and inevitably underperform in a competitive system (in which there must by definition be winners and losers) is more and more emphasised. Feelings of relative deprivation, low esteem and the existence of social division will increase, with all the attendant external costs, some of which have been described in the last chapter. The institution of 'league tables', whereby school exam results (usually GCSE and GCE A Level), are published in the press, is extremely divisive and deserves some attention here.

School league tables

Some say that the system of league tables for schools should be preserved, but put in the context of the social background of the intake. But besides the fact that it is absolute rather than relative success that would remain very influential on parental choices, exacerbating the existing trend towards social segregation, this ignores the fact that the league tables, however constituted, still serve to prioritise exam success as the sole marker for the success of the school and the individuals who attend it.

Even those on the centre-left who have attacked neo-liberalism seem blind to the corrosive effects of the school league table policy. Hutton, for example, comments: 'The league tables of school performance should reflect the catchment areas of the school, but their continuing publication is an important democratic gain and should not be withdrawn' (1996: 311).

Yet surely, the problem is that the league tables identify the catchment area of the school. Families want to send their pupils to the schools where they will mix with that peer group that is from the most opulent background. In that sense, parents are in fact acting rationally if their self-interest is gauged by what maximises the educational attainment of their children, for peer group influences are shown by research to be strong influences on the educational attainment of children. The league tables do not serve to highlight successful teaching: they serve to highlight social class background – and consequently what is happening is that the league tables have become very significant agents of social change. This social change is concerned with social polarisation. People decide to send their children to the schools which do best in the league tables: that is, have the highest levels of middle-class intake. Of course these schools become oversubscribed. They can expand their intake only up to a certain point, and they thus limit their intake to a geographically described catchment area. It is, in any case, difficult to send children to a school that is a long way from their home. So people try to 'buy in' to the geographical catchment area of the school by buying homes in that area. House prices in those areas go up. These areas become more and more gentrified. Other poorer areas around 'sink' schools, often attacked by OFSTED and stigmatised by poor showings in the league tables, experience plummeting house prices. The quality of life in these areas, which are increasingly stereotyped as ones being dominated by the underclass, goes down.

Far from improving educational standards, league tables are increasing social divisions that are already being widened by other techniques and practices associated with neo-liberalism. We have moved to a situation where the poor are trapped in certain areas because they cannot afford to move. They cannot therefore send their children to schools where the clientele come from more affluent families who, on average, have a more positive attitude to schooling and therefore a more positive peer group influence. In effect, the middle classes now send their children to what are really only quasi-comprehensives in leafy suburbs where the house prices are high.

It is ironic that the liberal centre-left, like Hutton, have spent so much time attacking private education and selection when they have for so long been ignorant of the social effects of 'parental choice' of schools and of school results league tables. Davies (1999: 4) has documented the trend towards social polarisation in the Sheffield constituency of David Blunkett, the Education Secretary of the UK Government. He describes the fate of two schools, now at opposite ends of the social spectrum in terms of intake, one of which used to be a grammar school, and the other a secondary school. He comments: 'Neither school is now comprehensive in anything but name. Neither school is any more comprehensive than it was 30 years ago. In those days, the children were selected by examiners. Now they are selected by estate agents.'

The financial arrangements governing the secondary sector produce a self-fulfilling cycle of decline for the schools in poor areas which lose pupils and therefore the grants that go with each pupil, while the middle-class schools gain money and the resources to produce better facilities. The richer schools are able to market themselves more successfully to the middle classes looking for areas in which to buy homes, that have the most attractive schools which their offspring can attend. In many ways the educational system is now different only in name from the old grammar/secondary-modern system, yet it parades itself as being egalitarian, and this, by conferring blame on the failure of individuals to compete successfully produces a socially and medically corrosive impact of relative deprivation. In the old days the working class may have been given inferior status and material rewards, but at least they had class solidarity to prevent individual blame. Now the social solidarity has been removed and the culture of blame falls on the individual. An individual who is so stigmatised by the system is not likely to want to co-operate or identify with that system.

The socially polarising effects of the competitive system in education are becoming better known, yet still there are few voices for the league table system, and other competitive techniques, to be abandoned. Even the critics seem dully resigned to the inevitability of the system, as if it represented some kind of progress that had its bad points but overall was moving us in the right direction. That there are no serious calls for league table and other techniques of competition in schools to be abandoned is a tribute to the strength of the neo-liberal discourse.

It would still be argued by many that the competitive system, as a whole, brings absolute gains, and that such gains outweigh any costs. But the notion that the system brings absolute gains to society, even before the external costs are taken into account, is far from proven. Indeed, it does seem that the application of competitive practices to schooling is an illustration that the injection of competitive arrangements into the provision of public goods can act to undermine the provision of the public goods themselves. In this instance, the market for provision of schooling is far from being perfectly competitive since the suppliers of education, the schools, cannot expand production beyond a certain point and take children beyond a geographical area because the number of school places is limited. Second, what is set up by the school league table system is not a market in schools according to their effectiveness in educating children but a market in the class nature of school intakes.

Co-operative solutions

As far as teachers are concerned, it is illogical to assume that increasing stress through ever more rigorous comparison, and in effect reducing their conditions of service, is going to increase the status of teachers so as to attract more recruits. Performance pay, even if it operates to reward those who are better teachers (something very difficult to define) will go to those who are 'better' anyway. While this will do nothing to improve the 'performance' of others it will demoralise these others by conferring upon them low status. Schoolteaching, which more than most professions, relies on co-operation to meet the needs and problems of individual children, can only suffer by bringing competition into the staff room. In fact, recruitment of teachers has, at the time of writing, reached a crisis point, so that rising numbers of children are taught by unqualified staff. Higher pay for teachers is no doubt part of the solution, but so is an end to the constant, institutionalised denigration of teachers. Instead, a culture of co-operation and mutual respect among teachers and between teachers, pupils and parents is one that needs to be fostered.

Of course, if conventional nostrums about improving the quality of children's learning are ineffective, and, in the case of the competition associated with the neo-liberal discourse, profoundly damaging, is there anything we can do to improve pupils' motivation to learn? The problem of low attainment is bound up with the way families at the perceived bottom of the social ladder see themselves and the small extent to which they often value education. What might make a difference, if the previous discourse on the benefits of social equality is accurate, is taking action to reduce inequalities of status and income. These two factors go together, and reducing inequalities of income may reduce the trend towards social polarisation where richer and poorer families are increasingly segregated from each other because of the widely differing ability to buy expensive housing. No doubt a greater equalisation of income and the ending of the school league tables' 'parental choice' policy would have the collateral benefits of reducing the destabilising effect of the cyclical housing booms, in part caused by a rush to grab the most desirable properties. 'Parental choice', of course, means choice for those who have the most money to buy property and the ghettoisation of the poor. Robinson comments:

> The really important conclusion to be drawn from the analysis of … [numeracy and literacy attainment data]… is that if social and economic disadvantage are so dominant in impeding basic skills development, then potentially the most powerful 'educational' policy might be one which tackles social and economic disadvantage.
>
> (1997: 17)

In other words, solutions lie primarily outside the education system itself. If the previous analysis is correct then, if we want to improve pupils' motivation to learn, we might be better advised to reduce social division, end the fetishisation of competition for material status and work towards making everybody feel they have a part to play in the system, which would be helped by the better-off acting as if they had a duty of care for others.

My analysis of Government justifications for increasing competition in education reveals that appeals to 'modernisation', 'progress' and the need to 'compete with other countries' are important elements in the neo-liberal discourse that legitimises the marketisation of society. That this definition of progress is widely accepted does not necessarily mean it is progressive or modern. The drive for improved performance is closely linked with a perceived need to compete with other countries. For a start, the most intense form of international competition – that is, warfare with other countries, as evidenced by the United Kingdom in World War II (Wilkinson 1996; Addison 1975) – is achieved in conditions of increased social solidarity and cohesion rather than in competition between individuals. Second, this notion ignores the fact that when people do things well it is generally because they value those activities, or at least are interested in them. Business consultants routinely give this advice to people hoping to set up in business, and warn that pursuit of money as the sole objective is likely to prove their undoing.

The road to better educational achievement may lie in pursuing learning as a good in itself, and it may be harmed and its morality distorted by posing it solely as a way of making money, or maximising rates of economic growth. Of course some comparison of levels of educational achievement is inevitable, for otherwise it would be impossible to measure people's differing skills for purposes of practical employment. I do not want to suggest that the division of labour be abolished, merely that the process be reconciled with the need for social solidarity. What is the gain in making a fetish out of competition? And why cannot other non-academic abilities of children be recognised and celebrated as a measure of status?

An early task in any social project for changing the values that dominate society at the moment would be the abandonment of competitive techniques of neo-liberalism such as performance-related pay and, in the United Kingdom, league tables of school performance. Given the experience of Switzerland, where the history of ethnic disputes has led to a decentralised political system and where there is no national curriculum, it is really very questionable whether anything more than minimal central control of curriculum ought to be wielded at the national or even state level. In Switzerland curriculum planning is left up to the schools and the cantons and the only national exams for university level are in French or German and mathematics. A more general alternative approach will be explored in the next chapter.

Having analysed the trend towards marketisation of society involving ever more frantic and fetishistic competition in pursuit of materialist objectives and the external costs for society that result from this trend, we are now in a position to flesh out more thoroughly the nature of the green theory that may provide a basis for a possible alternative to the neo-liberal discourse.

7
A Green Alternative

A culture of stress

As was described in the first chapter, the supremacy of the neo-liberal discourse in at least the Anglo-Saxon West rests, according to Galbraith's analysis, on the perceived self-interest of the contented middle-class majority of the electorate. This suggests that society is divided between the contented and an underclass who are alienated from the system that is seen to exist for the contented, but in which they see no point in engaging. The more unequal society becomes and the lower in status the underclass, the more they accept a parodied version of the neo-liberal discourse, that if the moral basis of society has been reduced to that of a competitive jungle and if the rules that do exist operate to oppress them and benefit others, there is therefore little point in respecting those rules. But this lack of respect is not expressed in a collective, class-based manner redolent of the labour movement of the past, but in an individualistic, hedonistic fashion. The lives of the underclass are, on average, much shorter than those enjoyed by the most contented in society.

However, perhaps one could say there are really three classes. The contented may be better analysed if divided into two parts: first, a 'super'-contented class, like the company directors whose 'performance' pay inexorably rises more quickly than anyone else's (and with little discernible connection with their performance). The neo-liberal discourse serves their interest well, for not only does it legitimise their increasingly large material returns and the otherwise insupportable inequalities that this brings, but it also increases their status and control. They live long lives. They are also increasingly separated from the rest of society by high-security systems inside which they vie among themselves to display their trappings of material success. More than 8 million US citizens already live in such enclaves, and this model is also being taken up in some parts of the developing world. Teresa Caldeira comments:

> In the context of increased fear of crime in which the poor are often associated with criminality, the upper classes fear contact and contamination, but they continue to depend on their servants....Public streets become spaces for the elite's circulation by car and for poor people's circulation by foot or public transportation. To walk on the public street is becoming a sign of class in many cities, an activity that the elite is abandoning. No longer using streets as spaces of sociability, the elite now want to prevent street life from entering their enclaves.
>
> (Caldeira 1996: 307–14)

Second, and by far the more numerous section of the so-called contented, are the 'wannabe'-contented middle classes. They aspire to the riches of the super-contented, and partly as a consequence agree, in the majority, with the notion that progress and modernisation are associated with increasing competition in pursuit of the maximisation of economic growth as the supreme objective. These wannabe-contenteds enjoy, by historical standards, high material living standards, though, paradoxically, are no happier, often less happier than their middle-class contemporaries earlier in the century. Yet they complain of increasing stress, induced to a great part by the very competitive systems that they regard as 'inevitable' parts of progress. They often lack control many feel the insecurity of short-term contracts, feel undervalued, fear the rising violence of the underclass, and aspire to the most opulent areas – involving often excessive mortgages which they are often barely able to afford but which they feel they cannot do without. Many of them may work in new fields, based on information technology, like banking and insurance tele-services, but in fact are subjected to the same if not higher stresses than those who belonged to traditional industrial trade unions. But they lack the protection and social solidarity that these working-class institutions used to involve.

These three classes are infused by a common drive to chase status through their consumption patterns (as described in Chapter 5, where the analyses of Baumann (1998) and Schor (1998) were discussed). This system enshrines relative deprivation as a necessary part of the system. A new conception of citizenship has emerged to give legitimacy for this new, neo-liberal order. Rather than the Marshallian notion of social citizenship which was given dominance through, for example (in the case of the United Kingdom) the Beveridge Report, we have a conception modelled on the Conservative notion of 'active citizenship'. This Tory notion, which emerged in the 1980s and is now accepted by New Labour, implies that the achievement of rights can only legitimately be gained by the acceptance of duties. Whereas in decades gone past there was at least the rhetorical Tory notion that the rich had a duty of care for the poor, nowadays the concept of active citizenship has a very different constellation of duties. Implicit in the neo-liberal discourse is the celebration of great inequality of rewards, necessary to give incentives, and the assignation of status through consumption. In effect, the rich have a duty to achieve conspicuous consumption in order to act as reference points for the rest, while the poor have a duty to re-educate themselves and make themselves suitably contrite in return for being allowed to be given welfare payments. There are growing calls in the United Kingdom for claimants to be made to sign on every day, and in the United States welfare payments hardly exist at all in some places.

Thus perhaps the dominant Anglo-Saxon political culture (and maybe, in the future, the German model) is at least as well described as a culture of stress as it is accurate to call it a culture of contentment. It depends on how self-interest is constructed and interpreted through discourse. And it is quite plausible to say that an alternative discourse, one focused on the social stress and external costs produced by an increasingly divided society, could resonate with the practical experiences of many of the so-called contented so as to displace the stranglehold currently enjoyed by the neo-liberal discourse.

It is certainly the case that there are plenty of critics of the socially destabilising effects of the marketisation process who operate from a conventional left or centre-left perspective. Will Hutton is perhaps the best-read and most articulate critic to emerge in the 1990s (Hutton 1996). Although his work was intended as a critique of the impact of Conservative policies, much of it still stands as a criticism of the Blair Government, which has accepted the main elements of the neo-liberal discourse. Although many of his criticisms will resonate with those made by the green movement, his proposals for institutional reform represent a top-down approach to reform state and corporate structures. This is rather than an effort to develop a practical critique of the competitive materialism that can reinvigorate society from the bottom upwards. Hutton denounces the growth of inequality and social exclusion as immoral, yet, apart from vague references to citizenship, proposes no clear moral basis upon which to promote an alternative vision. Hutton does not recognise the all-pervasive nature of what I have described as the culture of stress, and neither does he apppreciate how this culture of stress is based on a socially corrosive culture of consumerism.

A further crucial fault in Hutton's analysis, from a green point of view, is his complete, unexplained and rather curious omission of ecological considerations. Quite apart from the need to address the environmental issues, there is much in environmental and green thought that can be deployed to support a more morally based, co-operative and (in a governmental sense) interventionist approach. The Brundtland Report commented, for example, that 'if economic power and the benefits of trade were more equally distributed common [environmental] interests would be more generally recognised' (WCED 1987: 48). Greater equality leads to a greater will to tackle environmental problems, but there is also a good case for arguing that green moral philosophy offers an excellent basis for tackling problems of inequalities within human society. Moreover, it may offer a co-operative, less stressful alternative for the majority compared to neo-liberalism. Not only this, but the notion of external costs provides an excellent basis to model the human problems resulting from excessive competition. As I discussed towards the end of Chapter 4, environmentalists stress the existence of external costs of material production, costs associated with environmental degradation which are not included in commercial transactions and for which the polluter does not pay compensation.

But the external costs concept can be applied to human/human social relations when a particular type of production method, or social arrangement (such as excessive competition), harms people through stress, harm that is not reflected in commercial transactions. Just as greens deploy a holistic analysis to put the issue of external costs in the context of the relationship between humans and nature, so holistic concepts can be applied to put the issue of external costs of material production such as stress in the context of the relationships among humans in human society.

Holistic approaches to society

Hayward comments that

> Holistic metaphysics brings with it the idea that all phenomena are ultimately interconnected (versus atomism) and exist in one sphere

of being (versus dualism). These ideas have some far-reaching implications for thinking about the human relation to the rest of nature. They imply that the human mind is not a kind of entity which is radically different from other kinds of natural or material phenomena; they also imply that human societies are part of nature too and hence in principle amenable to naturalistic understanding. ... what makes ecology a subversive subject, some claim, is that it implies a view of nature and of human nature, individual and collective, which calls into question some of the cultural and economic premises widely accepted by western societies.

(1994: 30)

Although (as the quotation indicates) this description of holism is posited in the context of human relationships with nature, it is also very applicable to internal human nature and also, as is explicitly stated, human society. Re-read this quote again whilst omitting the second sentence. What Hayward says about human societies being amenable to naturalistic understanding implies not only a way of understanding the interrelationship between humans and nature but also a way of looking at the relationship between human individuals and human society. Individuals are part of a whole and the whole of society and the individual are mutually interdependent.

Of course, Hayward's formulation of holism merely expresses that which is common in the green movement, and indeed this sort of formulation has been put in the context of contemporary physics by Capra (1975), who cites contemporary quantum physical relationships to criticise the subject–object distinction inherent in classical Enlightenment science that appeared to set humans apart from nature. But what is less common is the application of the 'human in nature' holistic metaphor to the 'human individual in society' in the way that I have suggested. However, we can see how holistic metaphors have already crept out of strictly human–nature meanings and been applied to solely human endeavours when we look, for example, at the example of holistic medicine which aims to treat the whole body rather than just a part.

The 'human in society' holistic metaphor implies that the collective social good is dependent on individuals working together, and that the good of the individual is affected by the norms that regulate the collective. I have discussed in Chapter 5 how socially induced stress caused by relative deprivation resulting in social stress of various forms occurs when this moral imperative is perverted into one involving competitive consumption. In Chapter 3 there are examples of how this is being put into practice through techniques associated with neo-liberalism.

Towards the end of Chapter 4 I suggested a second application of the holistic metaphor to internal human nature, where humans cannot be reduced to machines operated by discrete chemical impulses but must be viewed as complex systems where psychology interacts with physiology. It is only in this way that the nature of stress and its effects on people can be studied and appreciated.

If we take these two human-orientated interpretations of holism together – the holism of internal human nature and the holism of the individual and society – and put them together with the holistic metaphor of humans in nature, we have the basis for a political project that is, indivisibly, both social democratic and green, or perhaps, if one defines socialism in terms of co-operative values, both socialist and green. However, rather than seeing green politics as a subset of socialist politics, a formulation that has been evident among eco-socialists, the (what might be called) 'social green' approach described here would interpret the social dimension as a subset of green politics. In western societies a social green approach sees the problems caused by social inequality as being bound up with relative deprivation rather than the left's traditional concerns with material deprivation.

Greens and socialism

While dissenting from Eckersley's 'intrinsic value in nature' ecocentric conclusions, I would concur with her comment that 'many eco-socialists regard the radical environmental movement... as part of a larger struggle to overcome capitalism' (Eckersley 1992: 127). Moreover, Marxism, despite some efforts to use early texts to argue otherwise, needs at least to be reformed, as Benton argues, in order to escape from the notion that in order to be free from necessity man must dominate nature. Humans must be in harmony with nature for their own interests.

A commitment to reduce competitive materialism does not necessarily mean we have to curtail most markets, as tends to be suggested by eco-socialists such as Ryle, who says that 'the planned meeting of social needs would be under direct social and political control' (Ryle 1988: 86). This is even more the case with Marxist eco-socialists such as Pepper, who argues: 'If ecological perspectives are to be combined with the traditional socialist desire to end capitalist accumulation and wage slavery (perhaps through the ultimate abolition of money) and to achieve equality, then it is clear that any eco-socialist society must be a highly planned and co-ordinated one' (1993: 444).

This leaves open the issue of who exactly is to be doing the planning and co-ordinating. A successful outcome of a Marxian class struggle – assuming the doubtful proposition that there were any traditional industrial workers left to mount this struggle – would leave us with fears that centrally planned production would end up entrenching the very industrially productionist forces that hamper efforts to achieve ecological sustainability. This is not to mention the general loss of contemporary faith in centralised planning to achieve socially desirable outcomes. Having said that, of course, in the United Kingdom industries such as railways and water where there is no competition and where the profit motive merely serves to increase costs, do seem to be candidates for a return to public ownership and to receive an injection of public funds to help achieve environmental objectives. But such decisions have to be made on pragmatic grounds rather than faith in some grand plan for centralised control of production.

Too much eco-socialist thought, then, is bound up with centralised concerns. Certainly we need intervention from government of various types, and practical policy ideas on how to reform structures, but a principal concern is to develop a new set of moral values and to find ways of energising the green movement to transmit these ideas throughout society.

Although, as we shall see later, Gorz also emphasised the need to plan for the production of socially essential needs, he was more perceptive in his analysis of materialism, which he associated with the determination of status and hierarchy: This analysis does well by describing the culture of consumerism, a consumerism that has taken an even more thorough grip on western culture than it had in the 1970s when Gorz wrote *Ecology As Politics*.

> Differences in consumption are often no more than the means through which the hierarchical nature of society is expressed. In extreme cases, the one and only purpose of distinctive consumption is to constitute others as poor, not to acquire anything that is intrinsically desirable. This is the case, for example, in the consumption of precious stones or high fashion articles. These conspicuous goods do not even procure pleasure, power or comfort: they simply demonstrate the power of acquiring things which are beyond the reach of others. The only function of these things is to make social inequality tangible.
> (Gorz 1980: 31–2)

A central problem with traditional socialist and especially Marxist thought, one that was observed by Gorz, is that it sees inequality as being synonymous with material inequality, rather than differential status. The left has thus concentrated on trying to advance claims for higher material returns, through trade-union action in support of high pay claims. Often the left, these days, are rebuffed when citing the exorbitant salaries of 'fat cats' as reasons for higher pay by saying that they are just playing the 'politics of envy'. The contemporary left often seem incapable of answering this charge. However, a green, less materialist approach might argue more effectively that material improvements for those in poverty (which in the United Kingdom usually means relative riches compared to the average person in the developing world) should be achieved in order to improve their status and self-esteem, preferably through employment, not merely to give them more cash to spend. Perhaps this is now an opportune moment to look at Gorz's alternative to classical Marxist politics.

Gorz's green socialism – a path to paradise?

Following Gorz's departure from the orthodox socialist body of thought, he developed a critique of capitalism, and the Marxist alternative. His ideas were still clearly influenced by Marxism; for example, through his emphasis on central planning of production of necessities, but he was also highly critical of Marxism and he recognised the key importance of the ecological crisis and the centrality of technological choices to political and social issues. He discussed, in *Ecology as Politics* (Gorz 1980), the idea that capitalism was generating vast social costs. These included not only ecological issues such as the destruction of natural habitats, the effects of road building and the impacts of nuclear power, but also various other social costs such as damage to health where medical institutions dealt with the symptoms, not the social causes, of ill-health. The key parts of his ideas on work, and his rejection of classic Marxist and socialist ideas, were rehearsed in *Farewell to the Working Class* (Gorz 1982).

Gorz's first task was to attack the notion, propagated by Marx, that the working class had the necessary autonomy to take over the means of production and organise it differently from the industrial productivism that is the hallmark of capitalism. Workers had become, or were becoming, mere slaves to the machine, mere functionaries, bereft of creativity and autonomy. In an analysis of power that has seemed influenced by Foucault, Gorz argues that power does not rest with the individual workers, but with the functions of the industrial system itself, which will carry on operating in much the same ecologically and socially malignant way whoever is in charge.

By its nature the proletariat is incapable of becoming the subject of power. If its representatives take over the machinery of domination deployed by capital, they will succeed only in producing the very same domination and, in their turn, become a functional bourgeoisie... .The notion that the domination of capital can be transferred to the proletariat and thereby 'collectivised' is as farcical as the ideal of making nuclear power stations 'democratic' by transferring their management to the control of the trade-union hierarchies.

(Gorz 1982: 64)

The second key point made by Gorz was that the current imperative of the industrial system is to destroy jobs through automation, thus producing a 'non-class' which at best enjoys marginal, part-time work. 'Technological development does not point towards a possible appropriation of social production by the producers. Instead it indicates further elimination of the social producer and continued marginalisation of socially necessary labour as a result of the computer revolution' (Gorz 1982: 72). Computerisation and automation of production was certain to carry this trend still further until jobs in manufacturing, and also jobs in service industries, were eliminated.

The answer to the social problems created by this process provides the third aspect of Gorz's strategy. This involves the collaboration of the movement (propelled forward by the new non-class) and the state to separate out a sphere of 'heteronomy' which would produce socially necessary commodities and a sphere of 'autonomy' where people engaged in liberating, creative forms of production and activity. The sphere of autonomy would gradually be extended while the sphere of heteronomy was reduced.

The extension of the sphere of autonomy is thus predicated upon a sphere of heteronomous production which, though industrialised, is restricted to socially necessary goods and services that cannot be supplied in an autonomous manner with the same efficacy. ... Heteronomous production may, for example, supply a limited range of sturdy, functional shoes and clothing with an optional usevalue, while an unlimited range of similar goods corresponding to individual tastes will be produced outside the market in communal workshops.

(Gorz 1982: 101–2)

The notion of 'convivial technology' is thus introduced, a concept discovered by Illich (1973). Later, Gorz defined autonomous production as 'essentially handicraft production in which the individual or the "convivial group" controls the means of production, the labour processes and the nature and quality of the product itself' (1985: 68).

Despite the reference to the 'market' in the earlier passage, Gorz emphasises the need for the socialisation of production and for central planning and co-ordination in the sphere of 'heteronomy'. The state, he argues, has a special capacity, which makes it the 'sole agency able to reduce socially necessary labour time to a minimum' (Gorz 1982: 115). Gorz supports Marx in the necessity for the socially planned production of needs as the precondition for the development of autonomy, but he stresses the need to demarcate this area of autonomy outside a circumscribed heteronomous area 'in which trivialised technical behaviour is the norm' (Gorz 1982: 114).

Gorz also discusses the notion of providing a guaranteed social income to everybody, rather than merely reserving welfare payments for the unemployed. He develops this point further in *Paths to Paradise* (1985):

> [I]n keeping with the socialist movement's original vision, the guarantee of an income for life is no longer seen as a compensation or allowance, or an extension of individual dependency on the state, but as *the social form which income takes* when automation has abolished, along with a permanent obligation, the law of value and wage labour itself.
>
> (1985: 42)

The influence, in France and in the EU as a whole, of Gorz's writings is extensive, and is most crucially exhibited in the drive towards legislative efforts to limit the length of the working week. There are many senses in which Gorz's work is resonant to themes explored in the last three chapters. His comments on the role of materialism in assigning status have been noted earlier. His criticism of the tendency of conventional work to restrict decision-making autonomy and, in a more general sense, that the current trajectory of capitalism involves wide-ranging costs, not merely ecological but also social in nature, chimes succinctly with the discourse on stress and excessive competition that has been articulated in this work.

However, there are also ways in which Gorz's ideas have weaknesses and vagaries. For example, Goldblatt (1996: 87) comments (citing

Giddens 1987): 'It has been argued that Gorz's claims for the impact of automation are overinflated, that his account of the structural transformation of the labour market is exaggerated, and that the political potentiality of the non-class of non-workers has not materialised.'

The political appeal of Gorz's proposals seems blunted by doubts concerning the apparent technological determinism involved in his notion that work as we know it is disappearing because of automation. Certainly, unemployment in the 1990s was higher than the 1950s and 1960s, but it was not higher than what was the case before what seems, on balance, to be this exceptional post-World War II period. Indeed, once one factors in the numbers of women who have come on to the labour market since the 1950s, what seems more remarkable is how the number of jobs has expanded to include them. In addition, as Schor (1992) has explained, a key social problem in the United States is that people at the end of the 1980s were actually working 20 per cent longer hours than they were at the end of the 1960s. Americans were working (and still work) over 30 per cent longer hours than Britons, and the British work considerably longer hours than the European average. Merely because manufacturing jobs, or even some service jobs, are disappearing does not mean that others will not be created. To borrow at least one neo-classical idea from J. S. Mill's *Principles of Political Economy*, if there are wants then people will go out and try and satisfy them. Moreover, the temper of the underclass, which broadly reflects Gorz's notion of a non-class of workers, is not, as the introduction to this chapter suggests, orientated towards collective political action.

A further area of criticism of Gorz is his sheer pessimism regarding the possibility of reconciling a degree of autonomy to what he calls 'heteronomous' production. The lack of autonomy, or decision latitude, is exactly what psychologists define as the prime source of work-related stress, but there is a growing literature on how that can be relieved by redesigning the working environment internal to the organisation, to which I shall refer later. How, anyway, is the state to decide the dividing line between heteronomous production of necessities and other items, and is the state really any better at doing that (or even as good as doing that) compared to the market?

A further issue is Gorz's use of Illich's notion of convivial technology. Gorz implies that this means low-tech handicraft communal workshops and the like, but Illich implies a wider usage. Illich specifically stated that the division between convivial and non-convivial tools was not about the level of technology. He cites telephones as examples of relatively hi-tech tools which fulfil the following criteria:

> Tools foster conviviality to the extent to which they can be easily used, by anybody, as often or as seldom as desired, for the accomplishment of a purpose chosen by the user. The use of such tools by one person does not restrain another from using them equally. They do not require previous certification of the user. Their existence does not impose any obligation to use them. They allow the user to express his meaning in action.
>
> (Illich 1973: 22)

The notion of convivial tools has parallels in Schumacher's (1973) notion of appropriate technology. He refers to a suitable piece of appropriate technology as something that is sufficiently cheap for the average person to buy, that is flexible in usage and that is reasonably compact. That both writers, operating from a decentralised, ecologically inspired viewpoint, and being highly critical of the technological, political and potentially socially centralising influences of nuclear power, should have similar degrees of support for decentralised technology is not surprising. A striking thought is that the personal computer, as used at home, now routinely connected to the Net, is an example of what could be the greatest convivial tool of our time. In fact, I have seen no evidence that this was in the minds of the two writers. At the beginning of the 1970s, when Illich and Schumacher published their works, personal computers were still a twinkle in Bill Gates's eye and the idea of the Internet would have seemed to belong to science fiction. However, the appeal of computer technology rested on the promise that computers represented both knowledge and therefore power. Gorz saw computer technology as something which reduced personal autonomy rather than one which enabled autonomy. At the end of the day, the reason that this technology took off was not because of some technological determinism, but because the technology happened to coincide with the cultural demand for more individualised access to knowledge and power from many individuals.

The bottom line of this critique of Gorz's work is that many of his concerns, and even parts of his solutions, are pointed in the right direction. However, we ought to be concentrating on the need for an expansion of personal autonomy and re-evaluation of conventional attitudes to work, materialism and consumerism on the basis of its being a good in itself, to help the health of the individual and society, rather than justifying it on the basis of some dubiously founded technological determinism about the 'end of work'. Let us look at some more morally based notions of how self-interest in the sphere of work may be refashioned.

The shape of future work

As far as conventional economic analysis is concerned, the key function of work is the production of goods and services. But Schumacher, borrowing from Buddhism, posited two other functions of work: to give the worker a chance to utilise and develop their faculties and also to enable them to overcome their egocentredness by joining with other people in a common task. Schumacher goes on to say:

> To organise work in such a manner that it becomes meaningless, boring, stultifying, or nerve-racking for the worker would be little short of criminal; it would indicate a greater concern with goods rather than people, an evil lack of compassion and a soul-destroying degree of attachment to the most primitive side of this worldly existence.
> (Schumacher 1993: 39–40)

This sort of approach is denounced as romanticist by many, who define their own approach as rational. Their interpretation of rationality sets economic objectives as the prime focus. But, as should be clear from the previous chapters, the lack of opportunity to use one's own skills and a lack of control over one's working conditions are major causes of job-related stress. If it takes Buddhist morality to highlight the plight of the expanding number of computer operators and telephone workers who act as conduits for impatient customers wanting to know their bank balances or the lowest quotes for car insurance, then this Buddhism has a lot to recommend. Clearly, defence of stressed workers striving to humanise their working conditions is, or should be, an important function of trade unions, and more will be said about this later.

Another 'Buddhist' notion – that of co-operation and team work – turns out to be, incidentally, highly compatible even with high levels of economic growth in the case of Japan. Co-operative methods are also much in evidence in Denmark and other Scandinavian countries which have very high standards of living.

Schumacher was not interested in promoting economic growth – he was a leading critic of the concept – but it is the case that ideas that are thought initially to be romantic can be accepted as rational (and even compatible with economic growth) if they are held to have some advantage. The process of deciding that they have some advantage is that of discourse, a topic that will be examined in the next chapter. But suffice it to say for now, ideas cannot be dismissed merely because they seem romantic or metaphysical. Even the Establishmentarian philosopher Karl Popper recognised this, in contrast to the views held by logical positivists who believe that only that which can be 'proven' is worthwhile. Popper comments: 'scientific discovery is impossible without faith in ideas which are of a purely speculative kind, and sometimes even quite hazy; a faith which is completely unwarranted from the point of view of science, and which, to that extent, is ' "metaphysical" ' (Popper, cited by Carvi 1997: 25).

Of course it is possible to look back at the writings of Schumacher and others, and on seeing that some of their ideas look out of step with what has happened since, declare that they were romantics. Such is the privilege of the self-styled rationalist whose only purpose is to justify what has happened and what is happening.

Robertson (1978, 1985) can lay claim not only to have absorbed eastern influences on the nature of work when he criticises the dominant rationality of work in the West, but also to having formed or to be forming, himself, a new, more ecologically benign work rationality. He discusses the evolution of the work ethic. He recounts how the work ethic evolved through Lutheran and Calvinist influences to mean that work was a means of obtaining salvation. He notes the similarity to and difference from the Marxist vision, saying that, 'whereas Christians perceive human work as a process of co-creation with God, Marx saw it as a process whereby human beings create themselves and, increasingly, the world around them' (Robertson 1985: 65).

Robertson discounts the leisure ethic as enunciated by Russell, who argued that work was to be avoided. He argues that people will resent the notion that they should merely amuse themselves without making any useful contribution, and that others would resent the idea of paying for their idleness. Instead, he sides with the distinction made by both Morris and Schumacher between 'good' work and 'bad' work. A 'new' work ethic is likely to evolve.

> Just as the Lutheran ethic taught that worldly work was more real than withdrawal into the artificial, abstracted sphere of ecclesiastical

life, so the new work ethic now will teach that to immerse oneself in today's organisational world is to sink into a world of abstractions and turn one's on real life; and that real life means real experience, and real work means finding ways of acting directly to meet needs – one's own, other people's and, increasingly, the survival needs of the natural world which supports us.

(Robertson 1985: 68–9)

Robertson's 'new' work ethic is desirable in the context of the holistic attitude discussed earlier in the chapter. It encompasses both relations between humans and relations between humans and external nature. It also offers a possibility for being a source of the spiritual dignity that is being lost in the shift from work ethics to consumer ethics that was discussed by Baumann (1998). Robertson sees what he calls 'ownwork' as bridging the gap created by the contraction of jobs in the manufacturing and service sectors. Robertson defines ownwork as 'activity which is purposeful and important, and which people organise and control for themselves. It may be either paid or unpaid. It is done by people as individuals and as household members; it is done by groups of people working together; and it is done by people, who live in a particular locality, working locally to meet local needs' (Robertson 1985: x). He sees the development of ownwork as a key part of what he calls the 'Sane, Humane, Ecological' (SHE) future. This will involve a greater amount of ownwork in place of full employment by flexible options: 'part time employment, self-employment, irregular and casual employment, co-operative and community work, voluntary work, do-it-yourself activities and productive leisure, as well as full time employment' (Robertson 1985: 131).

Like Gorz, Robertson argues for the introduction of a basic income scheme to break down the link between income and employment. This basic income would obviate the need for the complex system of benefits and tax relief now operating.

Robertson's ideas have a considerable amount in common with those of Gorz, but he does seem more in tune with the practicalities of economic and social life, and much less dogmatic about there being a centralising role of the state to divide up the economy into different spheres. For example, he discusses some practicalities of promoting community activity through local government and co-operative development agencies. Although Robertson also discusses the possibilities of technologically induced unemployment, he seems less determinist than Gorz in terms of the alleged inevitability of the 'end of work', and, all in all, Robertson presents a far less centralised version of developing the road towards greater autonomy. Robertson remains a radical green, but there are of course many ways in which the radical green questioning of the conventional work ethic has been converted into more reformist ideas.

Reforming work

One of the most popular writers on how people's work patterns can adjust to give them purpose, continuity and autonomy in their working lives is Handy (1995). He talks about how individuals can give themselves greater freedom by combining work and other activities, whether involving personal interests or voluntary community-orientated work. Handy uses the metaphor of a US-style 'doughnut' with a hole in the middle. The core, in the middle, provides the essential income sufficient to maintain the individual's preferred lifestyle, and the doughnut ring is the substance of the person's personal interests, or maybe short-term work which the individual does because of interest. Work, then, ceases to be involved with a single job (which, in practice, for most people does not provides absolute satisfaction), but is concerned with a portfolio of activities which includes both paid and unpaid activities. Handy argues that the portfolio will change according to the life cycle, with the core diminishing in size as people move into their late fifties or early sixties, or perhaps even earlier.

There is a significant branch of literature bearing upon stress reduction in conventional work situations. It is concerned with altering work situations so as to increase workers' sense of autonomy, or, to use Karasek and Theorell's notion of autonomy (1990), to increase their area of decision latitude. Two Swedish studies, one involving work at a Volvo plant and another at an old people's home, were used by the authors to demonstrate the benefits of a work environment where work was organised by employees in a co-operative and collaborative manner. Various other studies have analysed the benefits of similar schemes, including a group of studies documented by Kompier and Cooper (1999). These studies suggest that not only do co-operation among employees and efforts to increase the job autonomy of the employees help the individuals reduce their stress levels and improve their health, but also sickness rates are reduced and workers' motivation is increased. These latter factors produce increases in productivity. Liukkonen *et al.* (1999: 48) say that

> [A]nalyses ... have shown that the key financial figures, in particular the operational measures of effectivity, productivity and quality, are

inadequate for detecting early indications of conditions that may jeopardise a company's capacity in different areas. Instead it is customer satisfaction, working environment and health appraisals that clearly show when an organisation's capacity is insufficient.

Among the examples given (in the same volume as the quotation above) of successful antistress programmes is the case of the 'Hovedstadens Trafikselskab' (HT) bus company in Copenhagen (Netterstrom 1999). At the end of the 1970s it was found that sickness rates were 12–13 per cent and that there was a high turnover of bus drivers. Some bus drivers, interviewed independently of the trade union, came up with a number of factors that increased stress. These included the following points: that they were not involved in planning, that their complaints were ignored, and that they received little support from management when it came to illness and complaints from passengers. Driving time had increased, staff canteens were poor and the ergonomic design of the bus drivers' cabins was criticised. There were demands for more bus lanes and for ticket sales to be removed from the buses.

One route (166) was selected for a pilot study. A scheme was devised whereby the drivers were given a budget and were allowed to elect the service drivers. The service drivers acted as management. The rota was organised according to the wishes of the drivers, the drivers' cabins were redesigned and coffee machines and extra seats were fitted on the buses. Drivers were allowed to take their breaks at the bus garage where they could talk to others rather than at isolated places at the end of bus runs. New disciplinary procedures were agreed to combat lateness to work. In fact the budget was managed so as to achieve greater productivity, with the reduction in sickness rates allowing extra drivers to be hired and longer breaks to be given to the drivers. Passenger complaints also decreased. Netterstrom comments that 'The core of the project has been that the drivers have gained control over their working conditions through increased influence on the planning of the work, and through the power they have been delegated they have gained certain skills' (Netterstrom 1999: 192).

There was still need for drivers to be given training in how to deal with conflicts with passengers more effectively and how to act as team leaders. Unfortunately, due to company reorganisation which had nothing to do with the scheme, the route was put out for tender to another company and the maintenance of the scheme was not included as a condition of the acceptance of tender submissions.

Another example is represented by efforts to reduce stress at call centres, a topic discussed in Chapter 5. It has been suggested that in call centres 'giving individuals greater autonomy can reduce stress and staff turnover without affecting productivity' (Arkin 1997: 24). Greater autonomy means that call-centre workers are multiskilled, using their expertise to deal with a range of enquiries and being given greater ability to respond to customer's requests according to their judgements as opposed to merely parroting a pre-arranged script. Managers are more orientated to training and supporting their staff, and some experiments have been staged into operating with self-managed teams. But, according to the manager of a company that tries to reduce stress among its call-centre employees, 'this approach succeeds only where a degree of trust has developed between the agents and those who manage them. Without that trust, people will use greater autonomy as an excuse to avoid working' (Arkin 1997: 24).

Two points can be made in the light of these and other studies of stress reduction at the workplace. The first is that, while classic Marxist notions of workers' control merely means that the nature of the bosses changes but the alienation of the work remains, it is still possible to alter the way technology is used in conventional work situations so as to increase personal autonomy. Gorz is correct in pointing out that merely altering the structures that give the instructions (for example, workers' committees instead of boards of directors) does not alter the nature of the worker's job. In throwing out the Marxist version of workers' control, Gorz seems to overlook the possibilities for changing the nature of the job itself through extending the individuals' ability, through co-operation with workers and managers, to increase personal autonomy and reduce stress.

The second point is that noble though efforts to reduce stress at work through reorganisation of work environments may be, attempts to do so are likely to be short-lived and rare without a major change in the dominant discourse of neo-liberalism that sees employees as individualised units of production to be motivated only by financial inducements or penalties.

In other words, unless company attempts to produce low-stress environments are only going to be isolated oases, it seems that progress in stress reduction at work will proceed a lot faster if there is more general acceptance of a more holistic approach to work and health, a change in attitude which, in a green context, goes hand in hand with a change in attitude that favours more co-operation and less emphasis on consumerism. It could be argued that a social green approach reorientates attitudes to work and in a way which borrows from both eastern and western paths to holism.

Paths to holism: eastern and western

American radical ecologists such as Bookchin see hierarchy as being associated both with domination of other humans and human domination of nature (Bookchin 1980). Bookchin wants to see hierarchy kept to a minimum and decisions reached by co-operative agreement in decentralised communities. This can generate a social environment where sustainability can flourish and where humans have a chance to pursue a role of 'ecological stewardship'. It does seem, from practical studies of stress reduction at the workplace, that real (as opposed to superficial) empowerment can help to reduce stress and to increase people's sense of status and their own feelings of esteem. It can also give them a voice to improve the physical work conditions of their existence as workers. However, it also seems from the Japanese example that at the workplace at least (although not the home where male values predominate), a caring environment can be developed in the context of order and hierarchy. This works successfully when there is trust based on a common sense of endeavour, where workers and leaders have both rights and duties at all levels, with the leaders having a duty of care for their workers. This is in contrast to the individual competitiveness that is so widely promoted by neo-liberalism.

The sort of system that has developed in Japan means that company bosses are not seen to earn exorbitant salaries and they eat in the same canteens as the ordinary workers. Income differentials are quite low in comparison to the West, even though public spending is itself relatively low in comparison to some western states. In the West, especially in the United States, hierarchy is often seen as an opportunity to demonstrate status and superiority, and such differences are even held to be an incentive for greater effort. Economics theories with little or no empirical grounding, such as the 'Laffer curve', are adduced to suggest that the lower that the marginal tax rates for the rich become, the more incentive they have to go out and increase the GDP for everyone's benefit. Of course, such practices and ideas do not go unchallenged in the West, especially when writers coming from a green viewpoint are considered. Schumacher, commenting on an example of a well-organised, people-orientated firm, said that an important 'self denying ordinance' is that there should be a maximum spread between the highest and the lowest paid (before tax). He discusses a spread of one to seven (1979: 79). This sort of spread is very low compared to contemporary practices. A survey of top UK firms revealed that in many cases the highest paid directors received more than 200 times the average (never mind lowest) earnings of their workers (Abrams 1999).

In Japan the notion of rational self-interest is viewed differently from that which is dominant in the United States, and, also to a lesser degree, in Europe. In a Japanese organisation it is in one's self-interest to be co-operative so that the organisation can be competitive, not that (as in the West under neo-liberalism), the individual must be constantly competing with everybody else within the organisation. In a Japanese company an individual would usually do badly if they behaved in a way that was competitive as opposed to being co-operative with others. It is not that in Japan there are norms whereas in the United States there is just self-interest; it is that in the United States a different set of norms operate which shape self-interest into a different form. People follow self-interest not in terms of some constantly operating calculus, but in following roles.

Seeing self-interest as a co-operative venture does not mean putting aside the immediate self-interest of the individual in favour of some larger altruistic goal. The point is that if norms generally favour co-operation within an organisation, rather than competition, it is in the individual's own material short-term interest to be co-operative. Organisations will compete with each other, not the individuals. Such a co-operative strategy provides a suitable background for efforts to produce less stressful working environments, since the higher organisational aims (involving care for employees) as well as the individual welfare of the employee can be taken into account. The sort of holistic audit of company activities suggested earlier by Liukkonen will have a thorough grounding in organisational norms. However, if we view things atomistically, stress-reduction strategies at work are much harder to justify and usually emerge as no more than training schemes for methods by which people can cope with stress on an individual basis.

This brings us back to our earlier discussion of rational choice theory (RCT). The assumption (usually implicitly made by rational choice theorists) that people pursue a self-interest that maximises their own individual material returns is itself a structuralist assumption, because it is to assume a particular norm about how we maximise our self-interest. However, it is possible to have different normative structures that condition self-interest. As we have seen, in the east (Japan in particular) the norms for self-interest are concerned with acting co-operatively, and those who step outside this normative structure to act as competitive individuals will suffer. Their self-interest will not be served. Thus we can see that RCT can have very serious normative implications – and a normative wing of RCT is public choice theory – if it is implicitly conflated with the dominant western notion of individual competition rather than a norm of co-operation.

It may be expecting too much to ask for a sudden change in social norms in the United States and other western states such as the United Kingdom, but perhaps new organisational and wider social norms based on trust and co-operation could be fostered on a gradual basis. The West would do well to reap the benefits of co-operative lessons from the east while avoiding the dangers from unthinking obedience to hierarchy and the subjugation of women that is sometimes said to be a problem with eastern social systems. This would require a fusion of eastern concepts of the duties of leaders to care and the holistic nature of society, and also green and feminist ideas of decentralisation and self-determination on the basis of gender equality. Feminists writers such as Mellor criticise the way that green writers tend to see work purely in terms of alienated wage labour, and freedom as being liberation from this activity. Work, according to Mellor, is 'an integrated and obligated part of life which is much more women's experience' (1992: 244). In this conception, it is a matter of reorganising work, not of men becoming free from it and leaving the task of domestic work to the women.

Thus social green strategic changes in social norms and practices do not stop at the gates of work or school premises: they spread well beyond them, as the quote from Mellor emphasised earlier. The idea of reorganising life according to principles involved in what has become known as 'downshifting' can incorporate the nature and impact of these wider changes in strategic norms and practices.

Paths to 'downshifting'

In the United States authors such as Schor (1992, 1998) are evidence that there is a healthy subculture of scepticism concerning the legitimacy of the dominant competitive materialist culture. She discusses the trend, among some, to 'downshifting', involving working and spending less. She outlines nine principles to help people 'get off the consumer escalator... .I anchor my principles in the values of social equity and solidarity, environmental sustainability, financial security, and the need for more family and free time' (1998: 146–7).

Her first principle is that of controlling desire. People must be aware of the 'Diderot effect' (described in Chapter 5) whereby people buy one expensive item and then want many other expensive items to match it. People can curb habits of going through mail-order catalogues and of going to shopping malls as recreation that turn into bouts of comfort buying to relieve stress.

Her second principle is about making exclusivity uncool; that is, spreading a norm that shuns the idea of copying super-rich conspicuous consumption. I remember the motor vehicle sales advert saying 'You are what you drive', implying, for example, that if you are a high status person you buy a high status automobile. The structure of this preference is highly influenced, of course, by the company that is marketing the product, and it works because it resonates with a dominant materialist ethos. However, if identities are reshaped so that being environmentally sound is regarded as being of high status, then running around in a high-powered, gas-guzzling motor vehicle could be seen as a type of idiocy or boorishness.

A third principle involves voluntary restraints on spending, not merely on an individual basis but also on a community basis. An example of this could be attempting to combat the peer group pressure to spend more on clothes and expensive items that drives youngsters these days to spend so much time in part-time jobs just to keep up with their friends. Similarly, getting community agreements about limiting the sort of Christmas presents everybody buys their children can curb the upward pressures on spending that accompanies children's desires to keep up with the type of presents their peers are receiving. This can be related to the fourth principle, which is about using the community links to facilitate borrowing and lending items. Schor gives the example of the idea of having communal lawnmowers. Given that these machines are used for only a small portion of the time, the borrowing idea seems to make great sense.

The fifth principle is about becoming an educated consumer. This is partly about seeing past the designer labels to reach what we want out of a product rather than just attaining the status of having fashion chic, partly about knowing whether the good or service has been produced by sweated child labour, and partly by knowing the environmental impact of the process of production. It is also about knowing (and being taught) how to save. The sixth principle is about avoiding comfort spending, or 'retail therapy'. A large proportion of people do this regularly, and arguably just about everybody does it some time or other. A small proportion of people are said to be addicted. The seventh principle is about 'decommercialising rituals' such as holidays or buying presents for Christmas. Is it really necessary to buy expensive presents? Partly this is also about resisting peer group pressure. The eighth principle is about doing things like cleaning and travelling more cheaply, although the downside is that this way usually takes more time.

Principle nine is about co-ordinating responses, so that, for instance, tax rates on energy-inefficient cars are larger than taxes for relatively energy-efficient ones, and that expensive luxury items ought to be taxed more highly than others. Schor adds: 'Consumption taxes are a start. But mitigating the factors that give rise to competitive spending in the first place is also very important. This starts with reversing thirty years of growing inequality in the distribution of income and wealth.' If there was greater income equality, competitive spending pressures would ease, she argues.

It is desirable to illustrate this sort of 'downshifting' strategy in practice. To this end I interviewed a husband and wife team of downshifters.[1] I shall call them Jason and Margaret, although their real names are different. They live in the North West of England, and were at the time of writing in their late thirties. They both work part time and they share looking after their two children. Both of them are musicians in their spare time, Jason playing the trumpet and Margaret playing the clarinet. Margaret started job-sharing in 1992 following the birth of their first child (two years later another boy was born); she works as an administrative manager for the Probation Service.

> Although having the first child was an opportunity to go job share, I wanted to do it anyway as I was not enjoying the job so much. Financially speaking we really weren't much worse off compared to having to pay to keep the child in a nursery which was the alternative.

Then in 1995 Jason, who works as an architectural technician, fell ill suffering from fatigue and high blood pressure. He was off work for two spells of around three weeks.

> At first it was a question of cutting down the number of hours I worked so that I could cope with the job. I went onto half time working, although later I increased this to three days a week when both children started at school. But since then I have appreciated the advantages of working part time and I am not planning to go back full time, although it's possible I might reconsider when the kids grow up. Before I went part time I never knew how to stop. Now I have four day breaks in the week and I can unwind fully. I certainly feel I can relax a lot better.

I'm much more able to cope with the work and do a more thorough job. Now I like doing other things like walking, playing music and meditating. Doing housework during the day also frees up the weekends for family activities. You don't have to run around tearing your hair out doing household jobs and making the tea after you come home from work. We can share the housework and there's usually, except for half a day a week when the kids often go to their grandparents, one of us around to do it.

The good thing about being ill was that I've experienced bringing up children which is something I've enjoyed but is not traditionally done by men. It felt a bit odd at first going to collect the children from school or nursery when most of the parents there were women, but I've got used to that. I now would like to do this arrangement regardless of the illness – if it was a straight choice between full time and part time, I'd choose part time – but it's only because of the illness that the firm allowed me to go part time.

Margaret thinks that job sharing has helped her do the job better.

If I was doing my job full time I'd be off a lot more. Me and my job-share partner have very good sickness records and I feel a lot more positive about the job than I used. I certainly feel healthier at work. The main disadvantage of job sharing as a manager is that there's a fear that staff can play you off against each other, so you have to keep careful tabs on what has happened while you've been off.

The difficult side is the money. You have to cut your cloth. But we didn't buy an expensive house, so the mortgage isn't too burdensome. Not having to spend money on nursery fees was certainly a help in the past. But we still have to be careful what we spend. We will choose the cheaper options for holidays and when we go out we usually prepare sandwiches and a packed lunch rather than taking the kids to Macdonald's.

A different path from so-called 'downshifting' is outlined in books such as *Your Money or Your Life*, by Robin and Dominguez (1999, originally published in 1993). This involves accumulating savings sufficient to provide a modest income from the interest payable in respect of the savings. Dominguez grew up in a poor area of New York. He began earning a high salary and shifted to a new class of earners but he noticed that people were not any happier for having lots of money. So he decided that he wanted to use his time for his own purposes rather than using it up to pay for things which he did not need.

Blix and Heitmiller (1999) have produced a follow-up to *Your Money or Your Life*, a book inspired by their own personal transformation. The authors, who live together in Seattle, a place that has become quite a focus for green social and environmental values, used to be high-flying executives in the telecommunications industry based on the West Coast of the United States. 'We had luxury cars, nice vacations, a big house, we went wine tasting and all that. But it seemed like something was missing,' Jacqueline Blix explains.[2] 'I quit my job because it was becoming stressful. De-regulation of the telecommunications industry produced intense competition and great pressures for downsizing. This meant that the people who "survived" in their jobs were left with a lot more work.' David Heitmeiller was in a similar position in that he had five bosses in six years, and he would have to justify himself and explain what he was doing each time he had a new boss. 'Like me David realised that he was going nowhere. He couldn't even explain what his job was to his daughter. We just travelled a lot and went to lots of meetings.' The couple started to prepare for their downshifting in 1991 and then took the plunge three years later. Jacqueline had quit before David and had started a second degree followed by a master's and a doctorate in communications. She had some thoughts of going into public affairs but came to realise that it was essentially marketing. 'But I realised that it would be much the same as what I had lcft. I wanted to avoid another rat race.'

The couple heard of the 'New Road Map Foundation' inspired partly by Dwain Elgin's writings on voluntary simplicity and tried to apply the Foundation's nine steps to a new lifestyle.

> It begins by looking at your money. You think about what money you have made in your life, what you own, what you have achieved and how much time in your life went into achieving this. You come to realise you are spending an awful lot of time buying things that do not accord with your values. It simply is not a trade that is worth it. Becoming aware about your spending can be very revealing – some people spend $200 a month just on coffee – is this really worth the time you have to put in to raise the money to do this?

Now the couple live on $30,000 a year. This is hardly a starvation salary, but for two people it is well below what they used to earn in their former existence as yuppie executives. A lot of promotion of simple lifestyles is done through voluntary simplicity circles, a movement which has been given much publicity by the writer Cecile Andrews.

> At first we were just interested in the voluntary simplicity movement for the financial side, but then we realised the effect on the

environment that consumption has and also we have become much more attuned with giving things to the community and your family. For example, David spent a lot of time looking after his sick father before he died – so he now will not have the regrets that people have about not doing enough. He is also involved in *Habitat for Humanity*, the group that organises housing for low income families. We both now have a much stronger appreciation for us being citizens of the community and the planet.

It is a concern for the social dimension that makes us look at what distinctive approaches can be made by a social green alternative to the issue of urban regeneration – vital if social divisions are going to be reduced.

Greens and urban regeneration[3]

Some of the most innovative ideas on urban regeneration have come out of projects such as the Centre for Environmental Technology in Chicago or the London based New Economics Foundation (NEF), which have been inspired by ecological, de-centralist and co-operative values. Pat Conaty works for the NEF and has been active in organising initiatives such as local investment banks and credit unions in Birmingham, United Kingdom.

Conaty recounts the case of the South Shore Investment Bank in Chicago which has great metaphorical significance for a trend whereby even where, as is required by law in the US Mid West, banks have to be small and local in character, the savings of local people is not re-invested locally but goes out into the money markets where it attracts high rates of return. In the mid-1970s, in a campaign against financial racism the bank was bought out by a number of well-heeled charities, trusts and individuals for $2 million with the aim of using the $37 million of savings (which came from local people) for inward investment. Previously, practically all the investment went outwards and local people, predominantly Afro-Caribbean, could not succeed in application for mortgage loans. Since then the bank has invested in local businesses and housing, and the savings have increased to $600 million. Since the early 1980s there has been a big growth in community banking and credit unions in many of the major US cities, with about 50 now being well established. This is extremely small for a big nation, but pioneering examples are in evidence in cities like Boston, Philadelphia, Chicago, Los Angeles and San Francisco. Investment money comes from ethical investors, churches, general public, corporate responsibility funds and commercial loans. The banks, as a rule, recover around 80 per cent of their operating costs from commercial trading and around 20 per cent from grant subsidies from bodies like charitable trusts and foundations. The practice of outward investment into 'globalised' financial markets puts many downtown areas in a vicious circle of decline. The modest amounts of money that are available from a local community for investment goes out of the community, and does not come back in, thus reducing the local income and investment stream still further. Yet this practice is actively favoured by the neo-liberal discourse, which talks about how the system works best through competition to achieve the highest return, which is assumed to come from globalised markets.

Bigger banks tend to go for bigger and bigger investment partners thus making the cost of small customers very high. The worldview is one of 'biggest is best'. This creates a gap in the market for the reintroduction of community banking. The agency best placed to plug the gap between charity and increasingly globalised commerce is the co-operative movement. NGOs are most concerned with welfare and the state is increasingly centralised.

Mutual organisations and co-operatives are ideal agencies if it is actually possible to revive the local economy. The local economy is dumped on and regarded as history. It needs a level playing field. The trend is to marginalise the family business. We need a trend towards ecological taxation, less taxes on wages and more on resources and more import substitution as far as the local economies are concerned. We need incentives to encourage financial institutions to invest locally. Globalisation has been destroying local economies through franchising.

Local government could have a new role to play in urban regeneration. The peak period of local government in the United Kingdom as well as the United States was in the nineteenth century, the same time when co-operation was growing but after 1910 it was the corporations that took the lead. Since the Second World War we have seen a decline in local government and small businesses, although since the mid-1960s we have seen a revival in the small business sector. The regionally based, family-based and self-employed based businesses can promote local economic independence.

There is a powerful argument for subsidiarity both at a political level and at an economic level. Susidiarity was raised by Schumacher in *Good Work* (1979) and is about decisions being taken at their lowest practical level. Democracy shouldn't be a commodity that one buys every five years. Local councils have few powers these days and it is no wonder that fewer people are interested in participation. If you want to promote interest in local government, give local government more control over budgets, but let this control be exercised at the lowest possible level – at the neighbourhood council and parish council level – these levels should have budgets. Then people will be interested in talking about what should be done locally. Let's give a push for daily democracy. This will create an atmosphere more conducive to co-operation.

We need a new movement of co-operatives. My argument is not romantic or nostalgic. It is about how to put 'small is beautiful' ideas into practice. The mutual and co-operative sector could create a new culture of co-operation and participation in decision making which can feed into public affairs. We can be taking control of our own directions. There are examples of this happening in Italy, Spain and India.[3]

Putnam (1993) in his classic study of local political and social institutions in Italy showed that the best predictor of local government effectiveness was the degree of participation in local voluntary organisations of a range of types. Such co-operative, 'third sector' institutions, the sort of institutions that should be favoured for public support according to authors such as Rifkin (1995), engender social solidarity and trust among people – the same sort of factor cited for making co-operation work among call-centres workers described in Chapter 5. If politicians want to improve the voter turnout in elections, then, according to a social green perspective, they ought to think about ways of reviving the local political and economic culture rather than merely coming up with ideas like giving people the ability to vote in supermarkets.

Credit unions run mainly by volunteers are revivifying local environments where they are operating, particularly in the US. This sector is growing because of the way that the poor and relatively poor have been frozen out of the financial system. The state does put some funds into urban regeneration. Unfortunately the emphasis on urban regeneration is too much on the physical, housing provision side. That is often part of the measures needed, but more needs to done in terms of local involvement and development of local small business.

Even when that is attempted the whole process is short circuited because of deadlines imposed by central government which means that the programme organisers go for the 'quick fix' big projects which can defeat a major part of the idea of reviving the local economy when the spoils go to outside big companies who come in and do the orders.[3]

According to the social green approach, local democracy, participation and local economic revival go hand in hand. This is in contrast to how the neo-liberal discourse ends up, distorting efforts to help revive depressed areas. One hears government talk about injecting local economies with the zeal of competition and ensuring that local schemes are organised according to the best competitive practices. Yet if the super-competitive system with all its 'big is best' financial market tests has allowed the accumulation of depressed areas and the loss of social confidence, then there is a strong case for rebuilding social confidence and social capital in a way that avoids the excessive competition promoted by neo-liberal governmentality.

The projects that are funded by public money that work are small, community chest things like play schemes, community safety, environmental schemes like renewing canals, food co-operatives, community shops. They work because small amounts of money can go a long way. If you start off with big budgets you end up with the big outside providers queuing up. The problem is that the small schemes are not taken seriously. It takes time to build up the community confidence and skills necessary to manage budgets. Then, after a steep learning curve involving widespread experience and training, you can build up, in an organic fashion, to manage bigger budgets on a basis of well-developed systems of local participation. Projects that deliver high-quality training such as the Heatwise energy conservation, Wastewise and Naturewise projects in Glasgow can give people the confidence to gain mainstream jobs and to set up their own businesses providing local services.[3]

In fact, the money required for urban regeneration is not likely to stretch state budgets if it is used in ways that are efficient in social terms. The state can 'top up' the coffers of community banks to bridge the gap between its commercial returns and what might be obtained for investors by investing in the money markets instead. Budgets could be controlled locally and built up slowly, investing in businesses and services that flow entirely from local expertise rather than from colonising franchisers or franchisees.

Conaty claims that this sort of social green strategy which combines environmentalism with social justice is not a romantic return to nature but a serious strategy for inner city revival. But is there any basis for claiming that this strategy can offer an alternative path to progress to that offered by neo-liberalism and its vision of globalised markets providing social salvation; or is it as outdated as some of the green movement's detractors would like to suggest? I shall pursue this theme in the next chapter.

8
Truth, Technology and Progress

Fukuyama, whose writings have been widely accepted as emblematic in describing the alleged historic victory of capitalist liberal democracy, probably expresses the dominant view of neo-liberalism when he attacks the green movement as being infected with Rousseau-like romanticism. He implies that greens are against progress and technology. He says:

> His [Rousseau's] criticism of the Economic Man envisioned by John Locke and Adam Smith remains the basis of most present-day attacks on unlimited economic growth, and is the (oftentimes unconscious) intellectual basis for most contemporary environmentalism. ... Is it possible to imagine the emergence of a highly radicalised environmentalism that would seek to reject, on the basis of an updated Rousseauism, the entire modern project of the conquest of nature, as well as the technological civilisation that rests on it? The answer, for a variety of reasons, would appear to be no.
> (Fukuyama 1992: 84)

Fukuyama goes on to say that people will not return to nature and they will not freeze technology. I am sure they will not. However, the issue is not whether greens are Rousseau-ites any more than whether some right-wingers support General Pinochet's fascist methods because they support the introduction of market economics into Chile. Rather, it is whether unlimited economic growth is the main priority of social development, whether the current deployment of industrial and social technology (of extreme competition) is healthy and whether the conquest of nature really is a desirable philosophical goal. I would say that it is not that greens are against progress, it is a question of wanting a different sort of progress from the blind industrialism and extreme competition which seems to be favoured by Fukuyama.

This chapter is concerned with sketching out the philosophical basis for the sort of green response (as elaborated in the previous chapter) to the kind of criticisms made by Fukuyama. I am concerned with a response which can claim to be supportive of progress and which is not against technology. Rather, this response wants to see technology used for differing ends, ends that propose harmony with nature and which promote a political economy that avoids the external, stressful consequences of excessive competition and economic social division.

However, such a normatively favourable, albeit green, version of progress and technology also has, in the context of this work, to be reconciled with my discourse analytical approach. I am convinced that it is reconcilable, but given the attacks on discourse approaches as involving 'relativism' – which is said to exclude the possibility of talking about progress or objective truth – it is necessary to explain why such notions may not be foreign to an amended version of the discourse approach. Foucault seems to reject notions of progress. As Rorty puts it: 'Foucault's Nietzschean attitude towards the idea of epistemology is that there is nothing optimistic to say...we need to see him [Foucault] as trying to write history in a way which will destroy the notion of historical progress' (1986: 46). Rorty, for his part, favours a belief in social hope and support for science on pragmatic, although not philosophical, grounds (Rorty 1982: 161–2).

The Foucauldian discourse approach that I have discussed so far suggests that one cannot discover the truth outside of discourse, that instead society produces truth, and that within discourse no statement can explain the whole of the truth. This is not the same as relativism, a term used rather more often as one of abuse by those attacking non-positivist authors such as Foucault and Rorty who use a discourse approach, rather than a term used by these authors themselves. Relativism, the definition of which can vary from author to author, generally involves the notion that there is no absolute truth and that there is no reality outside discourse. In this form it is thus the opposite of what positivists argue; namely, that there is such a thing as scientific and technological progress involving the gradual uncovering of the truth about the real world, a real world that has a reflection in the form of our ideas about it.

Positivists, and realists – the latter may accept that some ideas may be discursively constructed but also accept some degree of association between ideas and materiality – often accuse followers of the discourse approach of being practically, or actually, nihilists. They conflate the criticism of notions of historical progress with the notion that it is impossible to distinguish between right and wrong. In the argument that follows I imply that this is at best simplistic, and in many ways it is wrong. In any event, those who are dubious about claims to absolute truth have a philosophical disposition that seems to be better developed for an age of political and cultural tolerance and democracy than those who exhibit considerable dogmatism about their truth claims, which is the impression that many of the realist critics of postmodernism seem to project. In this context Foucault's criticisms of the authoritarian potential of totalising creeds seem to perform an excellent democratic service.

Nevertheless, this still leaves open the issue of whether one can fashion and use criteria for assessing truth and progress that are applicable to differing groups of people at different times. To what extent is there objective truth between differing discourses? Foucault's work, at least, seems to offer little grounds for identifying any objective truth and progress from one period to another. Is it possible to talk of criteria for environmental progress beyond discussing a series of socially and historically specific criteria? This is a question regarding the theory of environmental knowledge, or environmental epistemology. It is an issue that I shall tackle now.

I shall start with a discussion of epistemology (the theory of knowledge) and ontology (the theory of being), in order to discuss what makes some environmental discourses appear more resonant than others. I shall look at the relevance of postmodernism's critique of progress. I shall then discuss whether we can speak of technological progress before discussing some of Beck's ideas on technological risk. Finally, I want to sketch out an alternative way (to that of Beck) of analysing the development of human efforts to control the threats to its environmental welfare. In so doing I hope to talk about a notion of progress that may be compatible with my use of discourse analysis.

Epistemology and ontology

The discussion above of issues ranging from the problems with neo-liberalism to the controversy in the United Kingdom over road building should cast doubt on the positivist notion that we can understand the world through finding 'true' ideas which are a perfect reflection of an at least potentially observable reality. The nature of the 'realities' is too hotly contested for that to be the case. However, this still leaves open the question of how far one can go in asserting that ideas are simply discursive constructions without being linked to any underlying reality. I believe that the relativist/realist dualism that is posed by some is an artificial one. I would much prefer to see attitudes towards the existence of common truths as a continuum with logical positivism at one end and relativism at the other end. The part of the continuum that seems to be relevant to a particular area of study seems often to be dependent on the proximity of relations of the subjects under discussion. The more distant the relations between those subjects become, then the more the relevant ontology and epistemology tends towards the relativist end of the spectrum. For example, the amount of common reality and common truths between a pair of identical twins brought up in the same household is very high, although never completely similar. The amount of common reality and truth between members of different species on planets at opposite ends of the universe is likely to be rather lower, and a much more limited sense of agreement on the nature of reality would lead us to accept an approach much closer to the relativist end of the spectrum. In natural science positivism seems to be useful, partly because the objects under examination do not interpret the worlds themselves, and partly because at any one time there is a high level of agreement on what variables should be, and can be, excluded. Of course positivism seems less useful even in natural science when one takes a long-term view, as Kuhn's writings imply. In social science, positivism has a much narrower appeal than in natural science, yet here in social science even postmodernists will concede that there is a considerable degree of cultural common ground of people from the same social location.

This still leaves us with the problem of finding epistemological and ontological criteria suitable for assessing progress in environmental policy, assuming that such as project is possible. What makes some environmental discourses appear more resonant than others? Let us prepare the ground to answer this question by looking at a view that is sceptical of realism.

The notion of 'species realism'

Rorty dismisses the claims of epistemology 'that all contributions to a given discourse are commensurable' (1980: 316), and places hope in hermeneutics which sees participants in discourse as 'persons whose paths through life have fallen together, united by civility rather than by a common goal, much less by common grounds' (1980: 318).

The implication is that there are no common rules of knowledge – in other words, no epistemology – that can determine what is truth and what is falsehood outside a given discourse. In addition, he suggests that only conversation links people engaged in a particular discourse. Postmodernists (along with Rorty, who disavows the term) suggest that we cannot conceive of a 'reality' outside of our ideas of it, and that what passes for common ideas of reality, as far as such commonality can be said to exist, can only be ascribed to a common cultural experience attributable to a particular time and place. Such common cultural experiences are only understandable, as Rorty suggests, through hermeneutical study. Nevertheless, the effect of Rorty's talk of common civility is to suggest that there is the possibility of truth within societies sharing certain assumptions. In dismissing the charge of 'relativism' Rorty comments that: ' "Relativism" is the view that every belief on a certain topic, or perhaps about *any* topic, is as good as any other. No one holds this view. ... The philosophers who get *called* "relativists" are those who say that the grounds for choosing between such opinions are less algorithmic than had been thought' (1982: 166).

Rorty and others contend that there is truth relative to a given group of individual observers who use a common discourse, and that there is the possibility of progress in pursuit of a set of goals that are commonly accepted by a group of people. We can extend this notion to talk about the basic shared perceptions and truths that are common to all humans at all times in place and time. Bevir (1999) moves in the direction of such a formulation, talking about an 'anthropocentric' notion of objectivity. However, this is not a term which I use here. Anthropocentrism may, in the critical terms suggested by Barry (1999), provide a sound basis for grounding a practical approach to policy questions revolving around human–nature relationships, but I would not use the term in the context of this discussion about epistemology since it gives the impression that humans are, in terms of access to knowledge, necessarily the centre of the universe.

I may concede that there are no common ground rules to determine truth in the universe as a whole. On the other hand, although humans are apparently vastly different in their cultures and understandings, they do share a set of perceptions of reality and also a set of knowledge-truths which is common for all human civilisations. This becomes apparent when we compare human knowledge to that of perceptions and knowledge enjoyed by other species. Each species can be seen to have a set of knowledge that is common to that species, and which, as such, is distinct from the set of knowledge understood by other species.

I would call this view 'species realism'. Thus if we are talking about knowledge as it relates to humans, we can talk of a human realism. However, this is a 'thin' realism, one that is restricted compared to what is commonly thought of as realism which tends to be extended, at least in the case of critical realism, to knowledge of social structures.[1] An understanding of the notion of species realism may lead us to see how, in environmental policy, some explanations, some statements and/or some discourses may have greater resonance than others. Of course, what does emerge as having greater resonance does so as a consequence of the existence of not only basic human truths that may be common to all civilisations, but also as a confluence of shifting eddies of cultural understandings. As we shall see later, these shifting eddies of cultural understandings interact with changing possibilities for technologies to produce different understandings of truth.

The notion that knowledge is not merely a human attribute, but an attribute which is possessed by non-human animals and which differs in character from species to species is a notion that is well understood by many of the supporters of the animal rights discourse. In criticising the way that Marx distinguished humans from the animal world, Benton comments that, on the one hand, 'Marx exaggerates both the fixity and limitedness in scope in the activity of other animals, and the flexibility and universality of scope of human activity upon the environment' (1988: 9) and, on the other hand, 'Each species of natural being has its own distinctive mode or pattern of interaction with nature – its own 'species life' (1998: 13).

A view that insists that there is such a thing as animal knowledge may be controversial among psychologists, where the balance of opinion still seems to regard humans both as the only beings enjoying a state of consciousness and also as the only beings having the ability to form concepts. However, there are strong arguments to suggest that this view is more a reflection of dominant prejudices, and of what is currently perceived as being human interests, than of scientific truth, whatever that may be. It could well be that this dominant discourse is already coming under sustained attack. Murray, whose research has found that the personality traits among chimpanzees are comparable in their diversity and complexity to that of humans (Murray 1998), comments that 'The study of matters such as animal consciousness and communication is seen through human eyes on the basis of human knowledge and human conceptions of consciousness and other issues'.[2] Indeed, what is apparent in the debate about consciousness is the relative lack of use of the concept by psychologists themselves and the conspicuous lack of agreement among philosophers (who are left to discuss the nature of consciousness) concerning the meaning of the concept (Gleitman *et al.* 1999: 337).

Considerable evidence can be presented to support the notion that, on the one hand, human concepts of reality and knowledge are not the only ways of conceptualising reality and knowledge, and, on the other, that there are commonalties in the perceptions and truths of humans throughout their civilisations which are distinct from the perceptions and truths enjoyed by other species.

The species-specific basis of perceptions of reality can be observed, for example, in the way that the kestrel can see into the ultraviolet light range, a range normally invisible to humans. This enables it to see the urine trails left by its prey, voles. Of course, humans have a very different attitude to urine compared to many other species. We can detect it less, certainly a lot less than, for example, dogs. Not only are the dogs' perceptions of urine very different from human perceptions, but also different truths are attached to them. To humans, urine has a negative connotation, whereas to dogs it reflects various positive social truths such as the denotion of ownership of territory. Sometimes the species realities and truths can be quite stark in their differences. In the case of humans, disturbance of a family home will be related to increased protection for vulnerable children; in the case of rabbits the mother will routinely respond to the home being disturbed by eating its babies. A human can drown in water, but a fish will asphyxiate on land.

Of course, it may be argued, in a manner that is redolent of the dismissal of 'primitive' cultures as being 'inferior' to industrial ones, that animal knowledge is inferior to that of human knowledge. The implicit reasoning here is that animal knowledge is not associated with the ability of humans to improve what we regard as material welfare. Measurement of intelligence through assessment of competitiveness or improvements in GDP is perhaps not entirely foreign to the neo-liberal discourse, but it would be a curious way of ascribing value to different species notions of knowledge. It is also said that, because animals cannot talk – or, to put it in Habermas' terms, they have no 'communicative rationality' – they do not have knowledge. They certainly do have a sense of 'instrumental rationality' – that of attempting to work on, and struggle with, nature in order to secure their own survival – and in this sense they do have their own species-specific common knowledge, their own 'instrumental rationality'. The notion that animals do not have communicative rationality is under attack. They certainly communicate, have a common discourse at some level. Some means of communication – for example, through smell – are demonstrably more sophisticated than our own abilities in that range. So when Habermas says that communicative rationality is superior to instrumental rationality he is giving precedence to a particular human pre-eminence and a pre-eminence in particular types of communicative rationality at that. There are suggestions that animals in one location may pass on to each other a common set of cultural habits that is different from that of animals in the same species in another location. The copying of what Dawkins would call 'memes' (cultural habits) is a process that occurs in animals as well as humans (Dawkins 1989). Experiments suggest that chimpanzees can communicate in some ways. It is indeed a curiosity that many of the conclusions about the way that human psychology works (for example, through classic and operant conditioning) have been reached as a result of experiments with animals, and yet we still claim that humans are very different, in a psychological sense, from animals!

It is becoming increasingly apparent that the possibilities for the existence of what we might call non-human intelligent species are very high, given the thousands of millions of stars in just one galaxy, and the large number of planets being discovered circling even a few of our closest stars. In such circumstances it is most unlikely that species will have exactly similar perceptual mechanisms and exactly similar common truth systems (for example, about reproduction) that we can observe in humans. They are likely to differ at least as much as between humans and other animal species on Earth, perhaps even more so. In other words, humans have a different set of what might be called perceptual realities and also a different set of truths compared to other species.

Human truths and environmental progress

It seems clear that if one moves from talking about the possibilities of commensurabilities between all types of knowledge in the universe towards talking about the commonalities of knowledge among one species – namely, humans – then one also moves towards a more restricted notion of the commonality of perceived realities and truths. We are in the realm of the 'thin realism' or 'species realism' which I have just discussed. Even if we are discussing humans from vastly different historical epochs there is an important degree of commonality of knowledge, certainly on environmental issues.

Associated with a common sense of reality is a common set of human truths. Human truths include the notion that babies are to be protected in even extreme circumstances, that the young should be taught to speak and walk at a young age, that eating certain materials is useless or poisonous, that it is possible to drown in water, be burned by fire and so forth. There are, of course, arguments between different schools of psychologists on the extent to which our perceptual mechanisms are innate or acquired. However, whatever the 'truth' of this debate, it is indisputable that humans throughout history have been blessed with a common set of senses and a common physiology. These senses and this physiology operate in ways which are very different from those of even our nearest species-relatives. These realities lead to the development of certain truths about what will do us harm and consequently will determine what possible discourses are more likely to have resonance and others less resonance.

Clearly, those arguments which convince people that death or ill-health is associated with a particular practice will lead them to evaluate the practice in a negative way, even though this negative evaluation may nevertheless be offset by other perceived positive (and here I mean useful) attributes of that activity or put aside in view of more pressing threats to life. Of course human physiology (a human-specific reality) dictates the parameters of what is, or is not, injurious to health. Progress, then, can be described as those activities, choices and processes that are perceived as leading to advances in human health and well-being. This search for improved health, this notion of progress, is driven by our physiological nature as the most basic of realities. It is this search which will lay down firm criteria for judging which environmental discourses have resonance and which do not.

Notwithstanding the existence of a set of human-specific truths, there is a very large potential area of discursive construction in environmental policy. Even when one takes for granted the nature of contemporary understandings based on our type of culture and technology, scientific uncertainty leaves considerable room for interest groups to select interpretations favourable to their own cause, which is illustrated by Liberatore in the cases of ozone depletion and the Chernobyl accident (1995). Moreover, even that which we believe to be certain can later be found to be erroneous.

Although I have sketched out some grounds to believe that we can talk of a common project of human progress in terms of advancing health, we should be wary of making too many claims as to what represents progress. At least some of the postmodernist critique of progress is still relevant. It is necessary to examine what parts of these critiques are relevant and which parts are less relevant.

Postmodernism and environmental practice

To Blaikie (1996), postmodernist environmentalist ideas have four principal strands. First they challenge metanarratives which claim to explain whole systems and define history; for example, religions or western paths of development which are analysed as being foisted on resistant southern populations. Second, science is socially negotiated rather than absolute, and is to be questioned when it comes into conflict with local, indigenous knowledge. Third, reality is socially constructed, and an account based on definite causal links is replaced with a series of narratives which have equal truth standing. Fourth, reality emphasises the idea that people should be able to speak for themselves rather than often remote 'experts' making pronouncements and judgements on their behalf.

Sachs (1992, 1993) has been one of the authors who have been associated with a postmodernist trend in analysis of environmental issues in the South. He and others, such as Shiva (1989), have denounced 'green globalism' in which western 'expert' knowledge and western environmental 'solutions' are imposed on the South in an effort to continue the root cause of the problem: the unsustainable nature of northern economic development. Indigenous people of the South are said to be the losers of this new imperialism, along with the environment that surrounds them. The political trends to which postmodernist environmentalists give implicit, if not explicit, support suggest that salvation lies in the struggles of these people to develop a sustainability which is based on working out how life can be made sustainable on a local level. Hence one can see the origin of the celebration of 'diversity' by postmodernists.

There are various instances where bizarre policies have resulted from the imposition of northern-based notions of what is meant by sustainable development. Conservation parks created by driving out the indigenous people who may have lived there sustainably with the wildlife for tens of thousands of years are cited by supporters of what is called neo-popular developmentalism as examples of centralised, northern-based, eco-imperialism (Blaikie 1996: 84).

There are examples which clearly suggest that sometimes northern-based expertise is inferior, even in its own terms, to that based upon local knowledge. Mainstream scientists have confessed that the Inuit have a better understanding of the numbers of bowhead whales in existence than they do. Similarly, there is recognition that Tibetan herdsmen know a lot more about crossbreeding yaks to improve milk production than conventional scientists (Edwards 1998). In African agriculture local traditions are being merged with western expertise to develop techniques based on an integration of tree and field crops in opposition to the monocultural paradigm imported by western experts (Funtowicz and Ravetz 1990: 22; Pacey 1990: 203). However, the postmodernist distrust of centralised expertise can be applied, in some instances, to the relationship of expert and local knowledge in the North as well. Funtowicz and Ravetz cite an example of the relevance of local northern knowledge in the apparently very expert field of radioactivity:

> A recent study by Brian Wynne of the University of Lancaster has shown how sheep farmers of Cumbria in England have a better understanding of the ecology of radioactive deposition than the official scientists. The farmers would not have made the assumption that radioactive caesium would leach away through their thin cover of acid moorland soil at the same rapid rate as through lowland pasture. Also they would have recognised that high ground lying directly downwind of a major reprocessing plant – the nearby Sellafield plant of British Nuclear Fuels – is liable to have a different deposition pattern from remote fields. Although they could not criticise the technically esoteric measurements made by the official scientists, they were fully competent to evaluate their methods and interpretations at every stage.
>
> (1990: 22)

Even within the scientific community itself there are examples of how centralised expert, 'hi-tech' knowledge is given widespread credence over that derived by scientists through their own experience.

For example, prior to the 1997 Kyoto Conference which was discussing international agreements to limit greenhouse gases, NASA announced that its satellites had detected no significant global warming since it started measuring at the end of the 1970s. Ground- and sea-based measurements garnered from instruments which had been used consistently for many years were attacked for their inferiority as low tech, and therefore unreliable, sources of information compared to satellites. It then emerged that there are gross irregularities in matching data from around 15 different satellites. These include unforeseen orbital degradations and inconsistencies in the comparison between different satellites that may have been calibrated differently, so that differing satellites may have been measuring temperatures at the same place but at different times (Jones *et al*. 1997; Wentz and Schabel 1998). The inaccurate satellite temperature change measurements were trumpeted by opponents of strong international agreements to cut carbon dioxide emissions as evidence of the unreliability of the climate change theory.

It was not as if this satellite technology failure was unprecedented. In the early 1980s NASA was pronouncing, on the basis of its satellite data (Litfin 1994: 98), that there was no significant thinning of the ozone layer right up until, in 1985, Joe Farman and the British Antarctic Survey published their measurements (using equipment designed in the 1920s) showing a drastic decline in concentrations of ozone over the southern pole. NASA admitted, belatedly, that its satellites had been calibrated to rule out 'bogus' measurements showing large declines in ozone concentrations. As Olson puts it, in explaining Quine's view: 'For Quine (1972), for example, what count as data depend on the total theory proclaimed to be data, and not, as the positivists thought, on the incorrigibility of directly sensed experience' (Olson 1986: 161). When a centralised expert body selects data, it is selecting it on the basis of a theory, not on the basis of privileged access to truth.

This over-reliance on 'hi-tech' methods is an example of what Marglin, cited by Saurin (1993: 56), calls the deference (inherent in the practices of modernity) given to what is called 'episteme' – the (epistemic) knowledge of the expert – over 'techne', the (technical) knowledge of the labourer. 'Episteme' claims to represent truth based on universal scientific knowledge, while 'techne' claims are based on first-hand knowledge and experience. I do not claim that meteorologists working with ground-based temperature recorders are akin to an indigenous tribesman, but clearly there is something in Marglin's distinction which can throw light on some of the imbalances of thought in western society itself, never mind westerners' treatment of the South. Hi-tech, centralised knowledge is not necessarily superior, and by this I mean useful, in providing environmental solutions, than lower-tech methods.

Neither is centralised expertise necessarily superior to the practical experience of people in the field, yet such techniques are inherent, for example, in the practices legitimated under the neo-liberal discourse. Despite neo-liberalism's claims to decentralisation it, in fact, prescribes certain types of management (such as the need for management consultants) and behaviour (performance-related pay) as being necessary to fulfil the ideals passed down from the canonical texts of Adam

Smith. Education provides an example of the deference given to 'epistemic' over the 'technical'. Schools in the United Kingdom have been made to obey changing edicts on what constitute good teaching, despite the view of many teaching practitioners that what is relevant in one locality is less relevant in another and that a style which suits one teacher does not necessarily suit another. In the 1970s, primary school teachers were condemned for being stuffy traditionalists who did not allow children to experiment and discover, whereas by the mid-1990s these same teachers, having adopted more 'progressive' techniques, were being condemned with equal vigour for their 'hippie' ideas and attacked for not espousing traditional methods and virtues. Which of these two experts judgements is right? Certainly, both of them cannot be right and, if they are not, then the legitimacy of expert centralised knowledge over local practitioner knowledge is seriously challenged.

Although the trend in teaching at all levels has been towards greater centralisation of knowledge on what constitutes good teaching, it seems that the ability of individual practitioners to pursue their own styles according to local conditions is largely a reflection on the relative political power (and place in the hierarchy of expertise) of the teachers concerned. Primary teachers' methods are the most rigorously controlled, secondary teachers' less so and university teachers' least controlled of all. In the university sector lecturers are not (as yet), within limits, blamed for the failures of their students. It is assumed that most of the outcome depends on the motivation of the students themselves. Similar judgements could also be applied to primary teachers, but are not, and the explanation for this difference may be bound up more with social acceptance of the expert as having ultimate status, a status which the poor primary teacher, for all of their wealth of experience, is not granted. In this case the placing of responsibility for 'failure' of individual pupils on primary and, to a slightly lesser extent, secondary teachers serves a valuable function for the neo-liberal discourse and the interests which support this discourse. The lost children's motivation, which flows from the social division wrought by neo-liberalism, is hidden from view by the condemnation poured upon teachers in 'failing' schools, schools which, as is rarely pointed out, always coincide with deprived, inner-city environments.

In addition to postmodernism's healthy disrespect for expertise, a further useful function is served by postmodernism's identification of the power of narratives or world views which are spread through discourse, as the main arbiters of change. As Gare comments: 'Most [environmental] improvements which have been made have been brought about by the pressure of direct action and not by regulations' (1995: 78). Perhaps the case is stated over-flamboyantly; not all action that sways public opinion is demonstrative involving sit-ins and underground tunnelling. It can involve writing letters to legislators or, as is most common in the United States, launching court cases. Moreover, regulations are often useful, indeed sometimes downright essential, given the sometimes opaque and technical nature of pollution abatement policy. Nevertheless, Gare's sense still remains relevant if it is meant to understand that the inspiration for tough regulations, or even schemes involving market-based emission trading schemes, derives usually from the popular pressure to do something to deal with the problem which, in turn, comes from efforts to change public opinion. This can be contrasted with the focus of rational choice analysts who take preferences for granted and concentrate on designing systems which will alter individual incentives. However, discussion of system design is fruitless unless there is a will to do something about the problem; an essential pre-requisite for this to happen is for public opinion to be mobilised.

Of course it is here that doubts begin to creep in about the complete relevance of the postmodernist message – in part, precisely because there does appear to be a message, itself an apparent contradiction in terms. But a bigger practical problem exists with the substance of that message. Experts and expertly designed systems are usually needed at some point and there are many (for a given period and technological understanding) generally applicable solutions for a world which is interdependent and which shares a great deal of technology.

Take the example of renewable energy technology. No matter whether people in developing worlds choose to be supplied by local units off the electricity grid, or by more centralised units on the grid (and this will depend on purpose and situation), it helps, indeed it is essential, to share knowledge from 'expert' sources about what is the most efficient design for a solar photovoltaic (solar electricity) unit. The same issues are involved in the cases of energy-efficient fridges, energy-efficient computers, energy-efficient motor vehicles and so on. All this technology is crucial to the task of deploying sustainable-energy strategies. These strategies are essential if we are to deal with what is a global problem, that of climate change, a problem which requires shared, expert study both of the problem itself and of the solutions. Saying that sustainability should be gained on a local basis is fine. However, the process of learning how to do this in a way that achieves both the environmental objectives and the most basic developmental needs that are demanded by the vast majority of people in developing countries requires global learning and some degree of global co-ordination. What can be said about the climate change problem can be said about all environmental problems to a greater or lesser degree. Conventional science and engineering may have caused many problems, but they can also offer many solutions.

The postmodernist claim to abhor all metanarratives is dubious. In the case of environmental policy, their emphasis on local sustainability and de-emphasis of centralised knowledge is itself a meta-narrative, and their inability to promote specific solutions is a crucial failing. As Gare comments, while postmodernists 'provide considerable insight into Western civilisation's orientation towards domination, they leave it crucially unclear how people should respond to the present situation' (Gare 1995: 97). Moreover, putting a halo around the actions of indigenous people can be misleading, given that on some occasions at least the environmental damage wrought by indigenous people can be considerable and unsustainable. The sad case of Easter Island is an example and a metaphor for human attitudes to the planet in general. The inhabitants, who arrived by boat, gradually chopped down the trees to enable food to be produced for an expanding population. However, after all the trees had been felled and food production peaked, famine set in. And, the islanders had destroyed their means of escape since they did not have the means to build boats.

I suggest that while postmodernism's criticisms of alleged progress are often justified, this should not be mistaken with abandoning all hope of the concept. We can talk about progress in environmental policy as that which improves health and survival and we should pragmatically apply both centralised hi-tech knowledge, and local knowledge gained from experience, in pursuit of this progress. This qualified notion of progress would suggest that an unqualified rejection of what is traditionally known as technological progress would be wrong. Yet there exists the allegation, sometimes given sustenance by greens, that ecologists are sceptical of technology in general. What should our attitude be to technology, both from a normative and an analytical point of view?

Technology and progress

Postmodernist notions of environmental policy involving diversity and support for local people struggling against centrally imposed ethics is reminiscent of Foucault's later ideas about how individuals can struggle to reconstitute their own identities as an act of resistance in the struggle for power. Can this sort of activity be described as a form of progress, a form of progress which postmodernists do not wish to acknowledge? There is even greater suspicion that Foucault was suppressing an implicit acceptance that technological development was occurring over time, as opposed to there being mere differences between technologies of equivalent standing in different periods. Foucault describes the changes in power relations as changes which flow from the development of technology of surveillance, a social technology that parallels the development of industrial technology. We can see today how the development of technology multiplies the possibilities both for the collection of knowledge and its dissemination. This makes greater surveillance and control over nature possible. However it also increases the possibilities for the use of surveillance to resist domination (by greater scrutiny of those who try to dominate) and also for nature to be defended. It was in his later works that Foucault tended to emphasise the ability of the subject to resist domination and influence the nature of their own subjectivity.

If we apply the set of truths associated with the 'species realism' concept explained earlier, we could say that technological progress can be measured by the extent to which it enhances basic human aims of health and survival. We can judge the degree of progress by improvements in the quality of life as measured, ultimately, by life expectancy. This at least is an incontrovertible measure of quality of life; as such it is better than others. Other measures tend to be culturally specific, while long life expectancy has, in human societies down the ages, been valued highly. Many other trappings of what constitutes quality of life may be contested in terms of value and certainly priority. The upward progression in life expectancy may not always be linear, but the general direction has been upwards. Conventionally, welfare is measured by increases in material production, but, as we saw in Chapter 5, above a certain level of material wealth, the link between life expectancy and relative incomes of people in different countries has broken down. What matters more to life expectancy, it seems, is the degree of social equality, with the more unequal societies being associated with lower life expectancies. Clearly improving some technologies, although not all technologies, can lead to an improvement in the quality of life.

Nevertheless, greens are often associated with opposition to technological progress. Dobson, for example, proclaims that 'Wholehearted acceptance of any form of technology disqualifies one from membership of the dark green canon' (1995: 96).[3] However, is this academic assessment of green attitudes to technology really accurate in the light of a careful analysis of both the theory of technology and the practical activity of the green movement? Many of the green movement's critics are keen to accept and project the stereotyped image that greens are against progress, as represented by advanced technology. Greens are compared to romantics, latter-day followers of Rousseau who venerate the 'noble savage'. Greens are pictured as people who want to take us back, in part at least, to an age of superstition.

With the possible exception of some groups of pagan green spiritualists, it is completely wrong to talk about contemporary greens as having the same attitude to nature as people before the Enlightenment. In those days people felt themselves to be powerless in the face of natural forces that they did not understand. Prayer to God or the gods was their only salvation in the face of calamity, whether famine, plague or some other disaster. The Enlightenment suggested that science and technology could be applied so that people could do something to take control over their destiny that was more effective than mere prayer. More recently we have seen how we have been threatened by our own technology. Yet, as Melzer puts it, 'The mainstream movements of environmentalism, antinuclearism, and others are simply the modern technological attitude now extended to the force of technology itself' (Melzer 1993: 311). In this way greens can be portrayed as being in the vanguard of progress, not as its enemy. This green vanguard of progress may promote different technologies to the industrial mainstream, but it chooses those technologies, including social technologies, which uphold the most fundamental of human truths concerned with the advancement of health and survival.

Barry Commoner, one of the gurus of the new environmentalism that sprang up, in mass movement form, in the 1960s, commented: 'The real question is not *whether* we should use our new (scientific) knowledge but *how* to use it' (1966: 29). Commoner was energetic in his attacks on neo-malthusians such as Hardin and Ehrlich who urged controls on population growth. It was not population, or even the growth of affluence, that had been the principal causes of the environmental crisis, but faulty or inappropriately used technology. Instead, environmentalists had to concentrate on making better technological choices (Commoner 1972).

In reality, despite their professed anti-technological leanings, even the most radical deep ecologists tend to suggest a path of alternative, not zero, technology. Permaculture, often touted as an alternative to monocultural agriculture is an agricultural technology itself and is a clear illustration of how radical ecologists still advocate technology, albeit of a differing type from that which is used conventionally. The

Australian Farms and Gardens Network, for example, describes permaculture as a

> holistic ecological approach to the design and development of human settlement. It takes into account food production, structures, technologies, energy, natural resources, landscape, animal systems, plant systems and social and economic structures. ... Permaculture draws upon traditional practices of earth stewardship and integrates this with appropriate modern technology.
>
> (www.squirrel.com.au)

Hence, dark greens may project an antitechnological image, but this is concerned with a misperception of the nature of technology, rather than their practice having a genuinely antitechnological orientation. Their alternative is clearly technological in origin, and this alternative comes with its own range of specialist expertise, training courses and design schema to boot. Schumacher, cited by most greens as an example of their approach to technology, talked about appropriate technology, not no technology. Dark green aversion to some recent technological innovations such as washing machines that dispense with domestic drudgery have attracted criticism from ecofeminists such as Mellor. She comments that 'Green criticisms of modern technology ignore the fact that a lot of the work it has relieved was traditionally working class or women's work' (1992: 241). As a single male who normally lives on his own I can also testify to the liberating nature of washing machines!

Implicit in the 'techno-sceptic' attitudes that dark greens are said to hold is the notion that we are part of a uniquely technological era which involves a complete break with the past. However, given humanity's distinctive feature as a toolmaker this assertion is very questionable. In the next section I shall argue against notions that, as Saurin puts it, environmental degradation is a 'normal and mundane feature of modernity' (1993: 62). I shall also argue against the idea that fundamental technological threats to human survival are a sole feature of modernity.[4]

Technology and risk

The notion that environmental degradation is uniquely associated with modern technology does seem to be common ground between postmodernists and dark greens, as well as other key analysts, like Beck. Although Beck does not call himself a postmodernist, he does suggest the idea that modernity presents us with unique threats of apocalypse which threatens the very basis of industrial society. As Blowers put it:

> Beck describes the present situation as one of reflexive modernisation. By this he means that Western civilisation has led to a transition from an industrial society to a risk society and with it there comes the confrontation with the self-destructive consequences which cannot be overcome by the system of industrial society.
>
> (Blowers 1997: 855)

Beck's 'Risk Society' thesis, which emerged at the time of the Chernobyl nuclear accident in 1986, neatly crystallised in intellectual form the fears that individuals were becoming more and more subject to risk in both social and ecological spheres. In Chapter 5, I referred to Beck in the context of how traditional social bonds were being dissolved and how individuals were being forced to take responsibility for their own destinies in a new competitive era. Beck was especially innovative in that he linked this trend with the notion of ecological risk. Put together, this means that individuals were not only facing greater social risk (for instance, through fragmented family background and a leaner welfare state), but also risk from environmental hazards over which the individual has little control. This ecological risk was technological in origin, and a new type of technological risk was in evidence. He continues:

> By risks I mean above all radioactivity, which completely evades human perceptive abilities, but also toxins and pollutants in the air, the water and foodstuffs, together with the accompanying short- and long-term effects on plants, animals and people.
>
> (Beck 1998: 22)

Although he concedes that risk is characteristic of the industrial era, he contends that modernity has spawned new hi-tech risks that are global and which, 'In the afflictions they produce they are no longer tied to their place of origin – the industrial plant. By their nature they endanger all forms of life on this planet' (Beck 1998: 22).

Beck said that the stench from sewers that has been a feature of industrial, and even to an extent medieval, cities since their inception was different from modern environmental threats in two respects. First, 'it is striking that hazards in those days assaulted the nose or the eyes..., while the risks of today typically escape perception'. Second,

'[I]n the past, the hazards could be traced back to an undersupply of hygienic technology. Today they have their basis in industrial overproduction.' Today's risks, Beck concludes, are also different in that they produce a systematic way of dealing with the hazards introduced by modernisation, which is called reflexive modernisation (1998: 21).

What all this indicates to me is not so much that there has been a qualitative change in the nature of risk, but more that there has been a change in the way we perceive and deal with technology and risk. A central problem lies with Beck's understanding of technology, which he associates with hi-tech machine technology. If technology is interpreted even a little more broadly, then it becomes much less clear that the ecological risk identified by Beck is only associated with modernity or industrial society.

Kass (1993: 4–5) uses a broad definition:

> Technology, in its full meaning, is the disposition rationally to order and predict and control everything feasible, in order to master fortune and spontaneity, violence and wildness, and to leave nothing to chance, all in the service of human benefit. Technology is thus understood as the disposition to rational mastery.

Technology is therefore to be seen in just about every human effort to organise the world, from early efforts to develop agricultural techniques and build roads to microchip technology and the Hubble telescope.

Given this sort of explanation of technology, one can think of a host of technological risks that pre-date industrial society, never mind the nuclear age. In particular, the sewerage problems to which Beck refers did have invisible consequences in the form of conduits for the spread of diseases, often in epidemic form. Such problems had technological origins, for the possibilities of the spread of diseases like cholera were greatly enhanced by the development of urban civilisation, itself a social technology that was made possible only by the steady development of agricultural technology. Sometimes epidemic diseases became pandemics, like the fourteenth-century Black Death which, spreading from China, decimated humanity and killed between a quarter and a third of Europe's population. These are phenomena that are, according to even contemporary notions, global in character. Indeed, in contrast to the fears (as yet) about nuclear holocausts, they actually materialised in practice. Such pandemics were spread by the agency of human technology. Rats, which initially caused the Black Death plague to be passed on to humans, flourished in the towns. Towns were a technology that had steadily expanded since the development of agriculture following the end of the last Ice Age 10,000 years ago. Urban living had expanded greatly in the Middle Ages. The disease was spread along trade routes and by ships, both important social and machine technologies which had become parts of the civilisation that had developed following the establishment and proliferation of human settlement.

Agricultural technology developed very considerably during the medieval period through the use of better ploughs, the harnessing of horses and the use of crop rotation. Industrial developments like these brought great gains in material productivity (and possibilities for a tremendous increase in human population) but also great risks. Increased vulnerability to epidemics and pandemics was only one type of risk. Droughts that might not have proved so dangerous to nomads who could move to new pastures could be extremely hazardous to people reliant on growing crops in particular locations. Monocultural agriculture is especially vulnerable to disease. The 'potato blight' famine which caused so much death and misery in Ireland of the 1840s is a testament to this technological vulnerability. The importance of having the right technology was emphasised in a direct sense in Greenland when the climate cooled in the middle of the fourteenth century. The hunting technology of the Inuit enabled them to flourish, while the Vikings eventually perished after the climatic changes made their arable farming techniques redundant and they were unable to adapt to new technologies.

So we can see that contemporary suggestions, such as that put forward by Beck, that technological risk is a phenomenon restricted to what we see as late modernity, or more generally that technology represents some new type of environmental threat, are perceptions, discourses about technology and environmental problems which are rooted in the modern historical period. However, an alternative view would suggest that it is the perceptions of environmental problems and human technology that have changed, not the existence of fundamental threats to our well-being which are associated with the nature of human technology. What has emerged in the last hundred years is the usage of the term 'technology,' and a self-awareness of increasingly sophisticated machine technology. This is allied to the growing dominance of specialist expertise in science and technology that has superseded the earlier tradition of craftsmen applying science in the context of their own experience. What has also changed is the method and ideology of control that humans have tried to apply over our environment. Indeed, the differing types of control that humans have attempted to apply to the environment may be a useful model for analysing technological change and environmental policy. I identify three ages of control.

Three ages of control

The first age of human efforts to control environmental problems is associated with the pre-industrial phase. Then problems like famine and disease were seen as the result of natural forces beyond human understanding, even though, in retrospect, many of these natural forces had arisen as a consequence of technological change. The only effective response to, or form of control over, the problems, whether they were disease, drought, flood or other great calamity, appeared to be an appeal to the supernatural. An example of such efforts to control the environment is given by von Storch and Stehr (1997). They recount the reaction to the poor summers and consequent bad harvests, by English and European society, in the 1315–19 period. Just as now, the authorities, which in those days were largely represented by the Christian Church, regarded the issue as a control problem. The Church advised people to atone for their sins, appease the wrath of God by prayer and give money to charity. This stratagem appeared to be successful, for better weather returned. In the West, at least, this 'supernatural appeal' type of control became increasingly centralised and institutionalised in style during the Middle Ages. The end of the Middle Ages is associated with a fundamental challenge to this form of control articulated in theoretical form by figures such as Bacon, Galileo and Descartes. This challenge became known as the Enlightenment. It was a precursor of industrialism and was about humanity attempting to take control, through science, of its environment.

As the industrial period wore on people began to recognise many problems, such as the spread of cholera, as being caused by inappropriate use of technology such as contamination of water supplies by sewerage. Consequently they sought to counter problems by attempting to impose human control over nature through perfecting technological responses like clean, piped, water and sewerage disposal at sea. These days faith in technology has reached such a stage where although we know that weather events such as storms are the result of the complexities of natural systems, weathermen will still be blamed for the damage if they do not give exactly the right weather predictions. We seem to expect technological 'fixes' even to natural events that are clearly beyond our control. At least the medieval Church was always in a 'win' situation, come rain or shine (literally), which is more than one can say for contemporary meteorologists!

In the West, by the end of the nineteenth century science had established itself as the new method of humans to control their environment. In this second age of control people became aware of technology. Although the new creed of science and technology did not, initially, claim to replace God in theory, it certainly claimed to be speaking for God in establishing its role as humanity's new means of controlling its environment. Ezrahi (1994: 31) quotes a nineteenth-century response, by an engineers' association, to a demand that they adopt a code of ethics like lawyers. The engineers said that 'since the actions of engineers were checked at every point by the immutable laws of God and nature there was no possibility of undetected fraud'.

It was in the latter half of the nineteenth century that a recognisable environmental movement arose in countries like the United States and the United Kingdom. However, the environmental movement was principally focused on preserving nature. Campaigns against industrial pollution of various sorts often made little headway. A study of health and environmental arguments in New England (USA) in the last three decades of the nineteenth century records how general campaigns against industrial pollution lost key battles with industrial interest groups who avoided having to adopt potentially expensive anti-pollution measures. Priority was given to tackling the spread of diseases, and the developing study of germs was used by industrial interest groups to concentrate spending on improving hygiene through better sewerage disposal and on the provision of clean water supplies rather than curbing industrial pollution (Cumbler 1995).

A similar story is evident from study of the battle over clean air in the United Kingdom. There had been various attempts to achieve smoke abatement through the promotion of parliamentary legislation in the first half of the nineteenth century. However, public admiration for steam engines as 'miraculous agents of civilisation' (Ashby and Anderson 1981: 5) and apparently more pressing concerns with the effects of uncontrolled urbanisation and the spread of cholera (1981: 15) sufficiently dulled public belief in the efficacy of action to allow the industrial lobbyists to prevail in their opposition to air pollution controls. Although the Alkali Acts, 1874, instituted the first real controls on certain industrial emissions it was not to be until after the Second World War that events such as the 1952 London smog triggered action through 'the ripening of public concern' (Ashby and Anderson 1981: 105). This led to the passage of the Clean Air Act 1956. The 1952 smog was not greatly worse than smogs in the nineteenth century, but what had changed, perhaps, is that what were seen as greater environmental threats to human health in the form of diseases had been countered to the extent that policies of public hygiene prevailed. Thus it was not so much the strength of industrial interest groups alone that prevailed in many battles over pollution control in the nineteenth century, but more that their standpoint resonated with dominant conceptions of social priorities of the time.

The post-Second World War period saw the inception of what now looks like the third age of control over the environment. This involves the use of science and technology to control the environmental effects of the human project of controlling nature that evolved during the second age of control. Two trends are worthy of emphasis here. First, as Commoner has observed, the post-war era coincided with a great expansion in the use of synthetic chemicals. Second, the great struggle during the 1950s over atmospheric testing of nuclear weapons evinced clear evidence that the tremendous destructive capacity of the (then) newly invented atomic weapons was being translated in practice into widespread, lethal, radioactive contamination. This is a struggle that has been much forgotten of late, overlain with the debate over pesticide use which was catapulted into prominence by Carson in 1962 (Carson 1962), and the emergence of the mass environmental movement at the end of the 1960s. However, there was a major turnaround from official statements in the early 1950s that atmospheric nuclear testing posed no significant risk to people to acceptance, by the end of the decade, that this was not true. This was followed by the freezing of atmospheric tests. These events provided a colossal and unprecedented example of supposedly authoritative, 'official' government-backed science being made to eat its words. These days, in the United Kingdom, people might mention the 'BSE crisis', which involved an embarrassing turnaround by government health scientists, as a reason for scepticism of statements made by 'official' scientists. Yet this suspicion of 'official' science has been developing for a long period, starting with the risks of nuclear contamination, and events like the BSE affair seem to be only the latest, although best remembered, layers.

There could, since the 1950s, be no mistaking the need to use science for a new purpose, with that of stopping what some began to see as an uncontrolled industrial monster destroying the very human servants (and the ecosystem upon whom they relied for survival) that it was created to help. It is especially the post-war development, especially nuclear development, of machine technology, allied to the apparent western success in meeting other basic requirements for human survival, such as hygiene, that coincided with the development of the modern environmental discourse. The basic human truth of increasing the prospects for health and survival (see the species realism discussed earlier) is now not merely interpreted, as it mainly was during the earlier part of the Industrial Revolution, as increasing material production, but also it now involves improving the quality of the environment.

This drive to improve human health and survival is likely, in the future, if the arguments in the earlier chapters are anything to go by, to be interpreted also to mean achieving lower stress levels at work. Neo-liberalism supports a set of social technologies that have been described in earlier chapters as having significant external costs for individuals and also for society as a whole. Just as machine or chemical technologies with major external costs for the environment have had to be curbed or altered, so does the excessively competitive techniques used by neo-liberalism. However, this will have to involve a basic re-evaluation of people's self-interests in work rather than merely developing stress management techniques to look after the symptoms.

The final chapter summarises and concludes upon some key arguments covered in this book.

9
Concluding Comments

My discussion of the existence of externalities affecting human health which are associated with excessive competition acts as a key part of the social green critique of neo-liberalism. Nevertheless, an argument doubting the potency of my critique could suggest that the discourse approach, which as used by Foucault appears to deny the possibility of identifying absolute truth through discourse, cannot be used to claim any objective superiority for a 'social green' as opposed to a 'neo-liberal' discourse. There are two robust responses to such an argument.

Discourse and truth

The first response is that the neo-liberal discourse does not, after the externalities are taken into account, stand up even on the basis of its own criteria. If there is a link between, on the one hand, excessive competition and the degree of social inequality and, on the other hand, shortened life expectancy, depression and heightened levels of crime, then these latter factors, these externalities, ought in theory to be given a monetary value. I say 'in theory', for the task of assigning monetary values to environmental problem is itself fraught with controversy, as we have seen in Chapter 2. Assigning a negative economic value to cover the impact of relative deprivation associated with social inequality will, at best, be an extremely contentious process. Nevertheless, if neo-liberalism is to be assessed on its own terms, such a judgement ought to be made. Judgements also ought to be made as to whether increasing competition in, for example, education, really does improve the GDP. The arguments in Chapter 6 suggest that its effects are negative. For example, techniques such as school league tables highlight the dominant social intakes of different schools and thus further add to the move towards geographical social polarisation as middle-class parents move to middle-class areas where there are schools with middle-class intakes.

The second response to those who might argue that there is no way of demonstrating the objective superiority of the 'social green' discourse is based upon what I describe, in Chapter 8, as my 'species realist' approach. This suggests, as an amendment to Foucauldian and postmodernist epistemology, that there are grounds for assessing the objective value of a course of action for a particular species according to the contribution of an action to the health and survival of members of that species. In the case of the neo-liberal discourse, it is precisely because it can be attacked for the external, stress-related costs of the excessive competition which are associated with the neo-liberal discourse that it can be the focus of an attack on objectivist grounds. Other discourses – for example, one based on the sort of social green approach discussed in Chapters 4 and 7 – can be supported on the basis that they minimise such external costs.

Social green strategies

Having said all this, we are still a long way from putting such ideas into practice. Certainly, any attempt to promote a social green strategy should not be confused with the sort of 'stress management' strategy that is being pursued in some companies these days. Of course it is better for employees if their jobs are organised so as to increase their personal decision-making autonomy, although even this is often beyond the cosmetic type of stress management 'coping' strategies that are promoted by management. Structural factors whereby the middle classes have better access to counselling could mean that the people who suffer most stress, the underclass and the workers in the jobs with the least variety, challenge and status, can be made to suffer relative deprivation by the very arrival of the social construction known as stress.

We need to reverse the trend towards class polarisation by reducing status differences. We are pursuing an ethic of ruthlessly driving down costs through individualistically based competition and consumerism. Instead a strategy of minimising income differentials, encouraging teamwork and increasing job control and company loyalty through avoiding rather than looking for opportunities to 'downsize', may reduce negative externalities associated with stress at work. Business could be organised so as to reduce differences in status between workers and management so that people can feel they are working for a common cause and that the management have a sense of duty towards the workers. A new social ethic which does not put such a premium on achieving status through materialism and which encourages job sharing and part-time working as a means of creating more individual space and sharing child-rearing can spread around jobs as well as reduce stress. Co-operative values, co-operative organisation and a locally managed injection of public funds into inner cities have been shown, in prototype schemes, to produce considerable benefits in reviving inner-city fortunes.

Green parties are, in theory, well placed to promote this strategy. As studies such as Rudig (1993) show the values of Green Party members already, to a great extent, reflect the social green values described in this work. However, again, as shown by Rudig *et al.* (1993), these values are usually not shown by ordinary Green Party voters who respond only to the salience of environmental issues. Yet do Green Parties really focus enough on the wider, social green approach? Do they, for instance, campaign for trade unions to concentrate more on improving work conditions rather than their traditional concern with improving wages and preserving wage differentials? In theoretical term green politics must be widened from being seen merely as involving a political economy of the commons. Of course the political economy of the commons is of vital, and ultimately supreme, importance. Yet unless green politics is seen as involving a wider political economy resting on a critique of the negative social (as well as environmental) costs of excessive competition and consumerism, it will not fulfil its potential. Sadly, also the neo-liberal discourse will be 'let off the hook' as its own construction of material self-interest will remain largely uncontested in what is seen as the central political terrain of economic and social political debate.

There have been some attempts, through NGO-based movements, to combine environmental issues with anti-poverty and social justice objectives. In the United Kingdom, The *Real World Coalition*, formed in 1996 by a diverse range of NGOs, although now inactive, was an example of this. In the United States the environmental justice movement is another example. Lois Gibbs comments that 'environmental justice is broader than just preserving the environment. When we fight for environmental justice we fight for our homes and families and struggle to end economic, social and political domination by the strong and the greedy' (Citizens' Clearinghouse for Hazardous Waste 1990: 2).

Another point worth making about a social green strategy is that it is at least as much an attempt to change people's values at work, home and as consumers as it is merely about demanding changes in government policy. The strength of capitalism results from its organic growth. A social green strategy does involve changes at the governmental level, including taking away the apparatus that embeds neo-liberal practices like performance related pay, but it cannot succeed without major changes in value structures that can persuade people to change the way in which they live.

The traditional left shares some of this agenda. However, according to green political economy, a crucial shortcoming of much left-wing thought is that it implicitly assumes that people's self-interest is concerned with maximising their material returns. Most people want more money, but the conditions under which they want more money, what they want to do with more money, and the trade-offs between more money and other, less material objectives, are not givens. What, for example, is the optimum balance between different types of work and other activities? Such sets of values are influenced by the dominant ethos of the time. The traditional Marxist left, with its emphasis on how the working class are defined in solely material terms does not, today, pose a serious challenge to the dominant neo-liberal interpretation of self-interest which is based purely on maximisation of material self-interest through competitive behaviour.

Reconstituting self-interest

I have already referred to the role of the neo-liberal discourse in constituting the perceived political and economic self-interests of the bulk of citizens in the United States and the United Kingdom. People's perceived identities have been transformed. Many used to see themselves as being class-based producers who contributed taxes for the common good, but now many see themselves as competitors and also as consumers whose status is maximised by the conspicuous nature of their consumption. This has a profound effect on what people see as their interests. As producers, they feel obliged to accept, if not overtly support, techniques of management such as short-term contracts and performance-related pay. As taxpayers they see that taxation must be reduced as much as possible to allow the private sector to supply more goods through increased competition, and also to give consumers greater freedom to spend their own money.

The discourse on green politics and stress developed in this book involves a re-interpretation of people's interests, or, in Foucauldian terms, a reconstitution of them as subjects. There is nothing 'natural', preordained or inevitable about the contemporary excessive emphasis on making people become competitors. The contemporary constitution of the subject as a competitive being is the product of the confluence, a fusion, of several interests and ideas. Emerging ideas that equality is achieved through competition, neo-classical economic ideas, the economic interests of the super-rich, the decline of traditional manufacturing interests and the increase in material well-being of the majority of citizens are among the interests and ideas that have reconstituted the contemporary subject. However, emerging ideas about the damaging stress-related consequences of social division and consumerism could reconstitute the subject.

Rational choice and self-interest

This discussion is interlinked with the critique of rational choice/public choice models of analysis theory that is set out in this work. I have also developed a critique of the neo-liberal discourse by extending the critique of public choice theory which is its close relative. A criticism that is common to rational choice (RCT) and public choice theory is that their predictions are based on a given, taken-for-granted notion of rationality, or at least that they have to be if they are to produce worthwhile conclusions. However, a key lesson from Foucault's work is that there exists not one rationality but a multiplicity of rationalities. Predictions made by RCT are thus limited to a particular set of assumptions sited in a particular rationality or discourse. The more normative predictions made by public choice theory and by neo-liberalism are sited only in a rationality that defines self-interest in a certain ultra-competitive, materialist way which takes little account of the externalities of the very competition that it seeks to fetishise.

Self-interest is not something which can be taken as a given if one is seriously to analyse how political outcomes occur. It may be no coincidence that RCT, which takes self-interest as given (and generally implicitly assumes it to be material in nature), has become a more dominant mode of analysis precisely in the period when neo-liberalism has itself become dominant. The problem is that while RCT is promoted as an analytical tool, there is often only a thin dividing line between an analytical device which assumes that something does happen in a certain way and a normative statement that something ought to happen in a certain way. RCT, which concentrates on methodological individualism, has flourished in a period when maximising individual choice has, as is implicit in public choice theory and as is explicit in neo-liberalism, become the dominant normative political economy of our time. An analysis which concentrates on how individuals maximise their own individual material returns has the danger of turning into a normative prescription that people *should* maximise their own individualistic material returns regardless of other considerations.

RCT, as demonstrated in Chapter 2, has problems when it comes to analysis of new environmental issues where the very conceptions of self-interest, and the world views which influence them, are subject to radical change. RCT is clearly best suited (if it is to make precise predictions concerning outcomes) to situations where self-interest is not only clearly quantifiable but also static in nature. Given these limits, RCT has a tendency to emphasise those economic motivations that most unarguably reflected in commercial (or pocket-book) transactions and those self-interests that are conventional. Non-economic conceptions of self-interest, changing economic conceptions of self-interest and radically different conceptions of self-interest in general will thus tend to be overlooked or downgraded in importance by rational choice analyses. It may be no surprise to see that rational choice analyses of the global warming problem celebrate the conventional view that solving the problem will be difficult. Maybe it will be, but a discourse approach could suggest, as hinted in Chapters 1 and 2, that a different scenario is at least plausible.

I would not want to dismiss completely the use of RCT as a mode of analysis. It does have purchase, especially in the context of institutional analysis. However, its boundaries of usefulness should be recognised in practice, and not just nominally. Rational choice analysts should avoid the practice of recognising the existence of criticisms of RCT in one text, and then proceeding as if they did not exist in another empirical case study. It does seem apparent that the very scepticism displayed by discourse theory (and by Foucault) towards claims to rationality, and the means by which discourse analysis can chart the changing nature of rationality, does offer to throw light on areas which RCT leaves either in total darkness or shrouded in the grey light of confusion. There can exist a multiplicity of shifting rationalities, and it is as least as important to analyse how those rationalities emerge and change as it is to study the decision choices and likely outcomes obtaining within a given rationality set.

Correcting Foucault

It is certainly the case that Foucault's methods are vulnerable to attacks that they lack a method of analysing agency and also that they seem to rule out the possibility of progress in an absolute sense. It does seem intuitively difficult to chart the development of environmental issues without tackling these two problems. I hope that Chapters 1 and 8, in particular, have helped to adapt Foucault's approach to dealing with environmental issues with respect to the issues of agency and progress. The notion of progress is itself made allowable by the 'thin realist' or 'species realist' conception that I have outlined in Chapter 8, and which involves a common set of human truths concerned with the objectives of pursuing human health and survival.

In Chapter 1 I discussed how the development of power through knowledge has gone hand-in-hand with technical and industrial developments. Foucault commented that we must 'cut off the kings head'. He thus implied that the regulation of the state was no longer dependent on coercion in the name of the sovereign, but on a regime of power/knowledge exercised through the mechanism of techniques of surveillance.

However, this surveillance can result in what environmentalists regard as positive outcomes. The pessimism of Foucault's thoughts on the effects of surveillance may be misplaced, or at least too one-sided. Surveillance goes in more than one direction, bottom up as well as top down, and the power of industry is limited by the growing, self-asserted knowledge of the consumer and the voter. Chapter 3 emphasises how the role of science is changing in society, with environmental issues being a key aspect of this change. In a sense we are all scientists now, or at least large sections of the population are not content to leave matters to those scientists that are either appointed by the government or whose existence is based on representing the interests of entrenched industrial interests. Environmental NGOs play a key role in this realignment of attitudes. The new environmental discourse whose emergence I have briefly described is not sceptical of, or opposed to, technology, so much as it wants technology to be directed towards new objectives. If the prolongation of a healthy existence is a fundamental truth of human self-directed activity, then it seems likely that the political technology of neo-liberalism will be challenged and the use of rational choice theory as a political science tool will be more limited and carefully judged.

Notes

1 Discourse, Power and Environmental Policy

1 Interview with John Whitelegg, 11/4/99.

2 Rational Choice Theory and Environmental Policy

1 Dowding (1991: 148) comments; 'It is difficult to show why particular individuals have the desires that they do. It is a task for psychologists or perhaps sociologists. Instead rational choice theory makes certain general assumptions about desires. It assumes utility maximisation and generally (though not always explicitly) assumes that a monetary value can be assigned to that utility.'
2 Complete privatisation is described by Eckersley (1993) as the free market solution. This is even more radical than (according to Eckersley) the public choice school approach which accepts, begrudgingly, that the state sometimes has a role to play, albeit in organising schemes that are as market based as possible through, for example, tradable emissions schemes. In practice, free marketeers such as Anderson and Leal (1991) accept that such state organised mechanisms are occasionally necessary – although one suspects hell would freeze over before they would accept a traditional regulatory approach! Like Ostrom, Eckersley concludes that mechanisms for solving environmental problems should be chosen according to their appropriateness to the situation rather than their ideological correctness. She notes that privatisation has the disadvantage of 'freezing' ownership of resources in a world that is constantly changing. She also criticises the free marketeers for their anthropocentric orientations, although this is a criticism that would also apply to Ostrom.

3 Science, Politics and Environmentalists

1 A version of a part of this chapter was published in the journal *Politics*, 19(2), (May 1999) under the title *Epistemic Communities and Environmental Groups*.
2 Environmental problems have been defined as involving 'any change of state in the physical environment which is brought about by human interference with the physical environment, and has effects which society deems unacceptable in the light of its shared norms' (Sloep and Dam-Mieras in Glasbergen and Blowers 1995: 42).
3 Personal communication with Richard Page, Greenpeace whaling campaigner, 12 March 1998.
4 Personal communication with Gunther Wuppman, German Greenpeace campaigner, 9 March 1998.

5 Health and Materialism

1 Calculated on the basis of figures supplied to me by the authors of the study.
2 Personal communication with M. Bobak of University College, London, 4/9/98.

7 A Green Alternative

1 Interview with 'Jason' and 'Margaret' (real names withheld by request) conducted on 29/8/1999.
2 Interview with Jacqueline Blix conducted on 6/11/1999.
3 This section is mostly based on an interview with Pat Conaty conducted on 2/11/1999. The quotation marks indicate direct quotes from him.

8 Truth, Technology and Progress

1 In fact much of the critical realist argument referred to above is similar to that of the 'constructivism' described by Adler (1997). A key difference is that critical realism involves a relationship between structure and agency, whereby 'All human agency occurs and acquires meaning only in relation to already preconstituted, and deeply structured, settings' (Hay 1995: 200). This discussion of structures perhaps accounts for its popularity on the academic left since it allows class structures to be analysed as 'real' structures and can incorporate the notion that the strategic terrain favours certain political strategies – for example, those of capitalists – more than others.
2 Interview with Lindsay Murray, 21/4/99.
3 Dobson associates the ascription 'dark green' with that of ecologism, which in turn is taken to involve support for ecocentric ideas, opposition to economic growth and support for decentralised political arrangements (see Chapter 4).
4 Of course, technically speaking, what dark greens say and what postmodernists say, or at least are supposed to say, is different. Dark greens are concerned to give their own version of ecological and technological truths, whilst postmodernists are concerned, in theory, to recognise the equal standing in truth of all narratives. Unfortunately the two strands seem to have become meshed.

Bibliography

Abrams, F. (1999) 'Pay Gap Grows in Fat Cat Britain', *Independent on Sunday*, 21 Nov., p. 5.
Acheson, D. (1998) *Inequalities in Health*, London: HMSO.
Addison, P. (1975) *The Road to 1945: British Politics and the Second World War*, London: Cape.
Adler, E. (1997) 'Seizing the Middle Ground: Constructivism in World Politics', *European Journal of International Relations* 3(3): 319–63.
Aitkenhead, D. (1998) 'Comment: On Examination, Chris Woodhead's OFSTED is a Very Expensive Flop', *Guardian*, 12 June 1998, p. 18.
Aldhous, P. and Swinbanks, D. (1991) 'Whaling Ban versus Science', *Nature* 351(6324).
Anderson, D. (1985) 'The Cost of Extra Inessential Car Traffic in Inner London', PRA Note 7, London: Greater London Council, Transportation and Development Department.
Anderson, T. and Leal, D. (1991) *Free Market Environmentalism*, San Francisco, CA: Pacific Research Institute for Public Policy.
Arkin, A. (1997) 'Hold the Production Line', *People Management*, 3(3), 6 Feb., pp. 22–7.
Ashby, E. and Anderson, M. (1981) *The Politics of Clean Air*, London: Clarendon Press.
Atomic Energy Commission (1962) *Civilian Nuclear Power: A Report to the President*, Washington, DC: Atomic Energy Commission; cited by R. Curtis and E. Hogan (1980) *Nuclear Lessons*, Wellingborough, Northants: Turnstone Press.
Bachrach, P. and Baratz, M. (1962) 'Two Faces of Power', *American Political Science Review* 56: 947–52.
Ball, S. (1981) *Beachside Comprehensive*, Cambridge: Cambridge University Press.
Barry, J. (1994) 'The Limits of the Shallow and the Deep: Green Politics, Philosophy, and Praxis', *Environmental Politics* 3(3): 369–94.
—— (1999) *Rethinking Green Politics*, London: Sage.
Baumann, Z. (1998) *Work, Consumerism and the New Poor*, Buckingham: Open University Press.
Beck, U. (1998) *Risk Society – Towards a New Modernity*, London: Sage.
Beckerman, W. (1994) 'Sustainable Development: Is It a Useful Concept?' *Environmental Values* 3(3): 191–210.
Benton, T. (1988) 'Humanism = Speciesism: Marx on Humans and Animals', *Radical Philosophy* 50: 3–18.
Bevir, M. (1999) 'Foucault, Power and Institutions', *Political Studies* XLVII: 345–59.
Birkes, D. and Firkin, L. (eds) (1989) *Common Property Resources; Ecology and Community-based Sustainable Development*, London: Belhaven.
Bishop, J., Formby, J. and Smith, W. (1989) 'International Comparisons of Income Equality', Working Paper 26, London: Luxembourg Income Group; cited by R. Wilkinson (1996) *Unhealthy Societies: The Afflictions of Inequality*, London: Routledge, p. 76.

Blaikie, P. (1996) 'Postmodernism and Global Environmental Change', *Global Environmental Change* 6(2): 81–5.
Blanchflower, D. and Oswald, A. (1999) *Well-being over Time in Britain and the USA*, Coventry: University of Warwick.
Blix, J. and Heitmiller, D. (1999) *Getting a Life: Strategies for Simple Living*, London: Penguin.
Blom-Hansen, J. (1997) 'A "New Institutional" Perspective on Policy Networks', *Public Administration* 75 (Winter): 669–93.
Blowers, A. (1997) 'Environmental Policy: Ecological Modernisation or the Risk Society?' *Urban Studies* 34(5–6): 845–71.
Brooks, G. (1998) 'Trends in Standards of Literacy in the United Kingdom, 1948–1996', Topic Issue 19, Slough, Berks: National Foundation for Educational Research.
Boardman, B. (1991) 'Ten Years Cold – A Decade of Fuel Poverty', Newcastle upon Tyne: Neighbourhood Energy Action.
—— (1997) *Domestic Equipment and Carbon Dioxide Emissions*, Oxford: University of Oxford Energy and Environmental Change Unit.
Bobak, M., Hertzman, C., Skodova, Z. and Marmot, M. (1998) 'Association between Psychosocial Factors at Work and Nonfatal Myocardial Infarction in a Population-based Case-control Study in Czech Men', *Epidemiology* 9(1): 43–7.
Bookchin, M. (1980) *Towards an Ecological Society*, Montreal: Black Rose Books.
Brynner, J. and Steedman, J. (1995) 'Difficulties with Basic Skills', Basic Skills Agency; cited by P. Robinson (1997) *Literacy, Numeracy and Economic Performance*, London: Centre for Economic Performance, LSE.
Burchell, B., Day, D., Hudson, M. *et al.* (1999) *Job Insecurity and Work Intensification*, Layerthorpe, Yorks: Joseph Rowntree Foundation/York Publishing Services.
Buchanan, J. M. (1986) *Liberty, Market and State*, New York: Wheatsheaf.
Caldeira, T. (1996) 'Fortified Enclaves: The New Urban Segregation', *Public Culture* 8: 303–28.
Callendar, G. (1938) 'The Artificial Production of Carbon Dioxide and Its Influence on Temperature', *Quarterly Journal of the Royal Meteorological Society* 64: 223–40.
Capra, F. (1975) *The Tao of Physics*, London: Wildwood House.
Carson, R. (1997) *Silent Spring*, New York: Fawcett Crest.
Carvel, J. (1998) 'OFSTED Inspectors "Unlikely" to Improve School Standards', *Guardian*, 6 May, p. 12.
Carvi, R. (1997) *An Introduction to the Thought of Karl Popper*, London: Routledge.
Chakravorty, U., Roumasser, J. and Kinping, T. (1997) 'Endogenous Substitution among Energy Resources and Global Warming', *Journal of Political Economy* 105(6): 1202–34.
Cherfas, J. (1989) *The Hunting of the Whale*, London: Penguin.
Christoff, P. (1996) 'Ecological Modernisation, Ecological Modernities', *Environmental Politics* 5(3): 476–500.
Citizens' Clearinghouse for Hazardous Waste (1990) *Everybody's Backyard* 8(1): 2; cited by D. Schlosberg (1999) 'Networks and Mobile Arrangements: Organisational Innovation in the US Environmental Justice Movement', *Environmental Politics* 8(2): 127.

Clammer, J. (1995) *Difference and Modernity, Social Theory and Contemporary Japanese Society*, London: Routledge & Kegan Paul.
Clarke, P. and Wilson, J. (1961) 'Incentive Systems: A Theory of Organisations', *Administrative Science Quarterly* 6: 129–66.
Cole, G. (1944) *A Century of Co-operation*, Manchester: Co-operative Union.
Commoner, B. (1966) *Science and Survival*, London: Victor Gollancz.
—— (1972) *The Closing Circle*, New York: Bantam.
Cumbler, J. (1995) 'Whatever Happened to Industrial Waste? Reform, Compromise, Science in the Nineteenth Century Southern New England', *Journal of Social History* 29(1): 149–72.
Curtis, R. and Hogan, E. (1980) *Nuclear Lessons*, Wellingborough, Northants: Turnstone Press.
Dahl, R. (1957) 'The Concept of Power', *Behavioural Science* 2: 201–5.
Daly, H. (1992) *Steady State Economics*, London: EarthScan.
Darier, E. (ed.) (1999) *Discourses of the Environment*, Oxford: Blackwell.
Daugbjerg, C. (1998a) 'Linking Policy Networks and Environmental Policies: Nitrate Policy Making in Denmark and Sweden 1970–1995'.
—— (1998b) *Policy Networks under Pressure: Pollution Control, Policy Reforms and the Power of Farmers*, Aldershot: Ashgate Publishing.
Davidson, A. (1986) 'Archaeology, Geneology, Ethics', in D. Cozens Hoy, *Foucault: A Critical Reader*, Oxford: Blackwell.
Davies, N. (1999) 'Bias that Killed the Dream of Equality', *Guardian*, 15 Sept., pp. 1–4.
Dawkins, R. (1989) *The Selfish Gene*, Oxford: Oxford University Press.
Demeritt, D. and Rothman, D. (1999) 'Figuring the Cost of Climate Change: An Assessment and Critique', *Environment and Planning A* 31: 389–408.
Department for Education and Employment (DFEE) (1996) *Learning to Compete*, Cmnd 3486, London: The Stationery Office.
—— (1997) *Excellence in Schools*, Cmnd 3681, London: The Stationery Office.
—— (1998) *Teachers – Meeting the Challenge of Change*, Cmnd 4164, London: The Stationery Office.
Department of Energy (1979) 'Energy Conservation: Scope for New Resources and Long Term Strategy', Energy Paper 33, London: HMSO.
Department of Environment (1990) *This Common Inheritance*, Cmnd 1200, London: HMSO.
Department for Environment, Transport and the Regions (1998a) *1998 Annual Report,* London: HMSO.
—— (1998b) *A New Deal for Transport*, Cmd 6836, London: HMSO.
—— Press Office (1998c) 'Meacher Welcomes BP's Initiative on Climate Change', 18 Sept., London: DETR.
Department of Trade and Industry (1993) *The Prospects for Coal: Conclusions of the Government's Coal Review*, Cmnd 2235, London: HMSO.
Department of Trade and Industry and the Scottish Office (1995) 'The Prospects for Nuclear Power in the UK', CM 2860, London: HMSO.
—— (1999) 'New and Renewable Energy – Prospects for the 21st Century', London: HMSO.
Department of Transport (1989) 'Roads for Prosperity', Cmd 693, London: HMSO.
—— (1996) 'Transport, the Way Forward', Cmd 3234, London: HMSO.

Devall, B. and Sessions, G. (1985) *Deep Ecology*, Salt Lake City, Utah: Peregrine Smith.
Dews, P. (1989) 'The Return of the Subject in Late Foucault', *Radical Philosophy* 51: 37–41.
Dobson, A. (1995) *Green Political Thought*, London: Routledge.
Doherty, B. (1998) 'Opposition to Road-building', *Parliamentary Affairs* 51(3): 370–83.
Dowding, K. (1991) *Rational Choice and Political Power*, Aldershot: Edward Elgar.
—— (1995) 'Model of Metaphor? A Critical Review of the Policy Network Approach', *Political Studies* XLIII(1): 136–58.
Downs, A. (1976) *Urban Problems and Prospects*, Chicago: Rand McNally College Publishing Company.
Dryzek, J. (1996) 'Foundations for Environmental Political Economy', *New Political Economy* 1(1): 27–40.
—— (1997) *Politics of the Earth – Environmental Discourses*, Oxford: Oxford University Press.
Dudley, G. and Richardson, J. (1996) 'Promiscuous and Celibate Ministerial Styles: Policy Change, Policy Networks and British Roads Policy', *Parliamentary Affairs* 49:566–79.
—— (1998) 'Arenas without Rules and the Policy Change Process: Outsider Groups and British Roads Policy', *Political Studies* 46(4): 727–47.
Dunlap, R. (1998) 'Lay Perceptions of Global Risk: Public Views of Global Warming in Cross-National Context', *International Sociology* 13(4): 473–98.
Durkheim, E. (1961) *Moral Education*, New York: Free Press of Glencoe; cited by S. Ball (1981) *Beachside Comprehensive*, Cambridge: Cambridge University Press.
EarthWatch Institute (1998) *EarthWatch 1998 Annual Report*, Oxford: Earth-Watch Institute.
Eckersley, R. (1992) *Environmentalism and Political Theory*, London: UCL Press.
—— (1993) 'Free Market Environmentalism, Friend or Foe?' *Environmental Politics* 2(1): 1–19.
Edwards, R. (1998) 'The Price of Arrogance', *New Scientist* 160(2156): 18–19.
Ehrlich, P. and Ehrlich, A. (1990) *The Population Explosion*, London: Hutchinson.
Elliot, D. (1978) *The Politics of Nuclear Power*, London: Pluto Press.
Elster, J. (ed.) (1986) *Rational Choice*, Oxford: Blackwell.
Eyre, N., Downing, T., Hoekkstra, R., Rennings, K. and Tol, R. (1997) *Global Warming Damages*, Brussels: ExternE Project Report, Joule Programme, Brussels: European Commission.
Ezrahi, Y. (1994) 'Technology and the Illusion of the Escape from Politics', pp. 29–37, in Ezrahi, Y., Mendelsohn, E. and Segal, H., *Technology, Pessimism and Postmodernism*, Boston: Kluwer Academic Press.
Fleming, J. (1998) *Historical Perspectives on Climate Change*, New York: Oxford University Press.
Flood, M. (1985) *Energy Without End*, London: Friends of the Earth.
Foley, G. (1987) *The Energy Question*, London: Penguin.
Foucault, M. (1971) *Madness and Civilisation*, New York: Pantheon Books.
—— (1973) *The Birth of the Clinic*, New York: Pantheon Books.
—— (1993) *The Birth of the Clinic*, London: Routledge.

Foucault, M. (1980) *Power/Knowledge – Selected Interviews and Other Writings 1972–77*, ed. Colin Gordon, Brighton: Harvester Press.
—— (1977) *Discipline and Punish*, London: Allen Lane.
—— (1981) *The History of Sexuality, Volume 1*, Harmondsworth, Middlesex: Penguin.
—— (1991) 'Governmentality', in Burchell *et al.* (eds), *The Foucault Effect*, Hemel-Hempstead: Harvester Wheatsheaf.
—— (1994) *The Archaeology of Knowledge*, London: Routledge.
Fowles, R. and Merva, M. (1996) 'Wage Inequality and Criminal Activity: An Extreme Bounds Analysis for the United States, 1975–90', *Criminology* 34(2): 163–82.
Fox, W. (1990) *Towards a Transpersonal Ecology: Developing New Foundations for Environmentalism*, Boston: Shambala.
Freedland, J. (1999) 'Goodbye to the Oracle', *Guardian*, 9 June 1999, p. 19.
Friedman, M. (1982) *Capitalism and Freedom*, Chicago, IL: University of Chicago Press.
Fukuyama, F. (1992) *The End of History and the Last Man*, London: Penguin.
Funtowicz, S. and Ravetz, J. (1990) 'Post-normal Science: A New Science for New Times', *Scientific European* (Oct.), pp. 20–2.
Furedi, F. (1999) 'Time to Put Up and Shut Up', *Independent*, Real Life Supplement, 10 Oct. 1999, p. 11.
Galbraith, J. K. (1992) *The Culture of Contentment*, London: Sinclair-Stevenson.
—— (1999) *The Afffluent Society*, London: Penguin.
Gamble, A. (1994) *The Free Economy and the Strong State*, London: Macmillan.
Gare, A. (1995) *Postmodernism and the Environmental Crisis*, London: Routledge.
Giddens, A. (1987) *Social Theory and Modern Sociology*, Cambridge: Polity.
Gleitman, A., Fridlund, A. and Reisberg, D. (1999) *Psychology*, New York: W. W. Norton.
Goldblatt, D. (1996) *Social Theory and the Environment*, London: Polity.
Goodin, R. (1992) *Green Political Theory*, Cambridge: Polity.
Gorz, A. (1980) *Ecology as Politics*, Boston: South End Press.
—— (1982) *Farewell to the Working Class*, London: Pluto.
—— (1985) *Paths to Paradise*, London: Pluto.
Grant, W. (1995) *Pressure Groups, Politics and Democracy in Britain*, Hemel Hempstead, Herts: Harvester Wheatsheaf.
Green, D. and Shapiro, I. (1994) *Pathologies of Rational Choice Theory*, New Haven and London: Yale University Press.
Greenpeace (1994) *Supermarket Greenfreeze, Supermarket Refrigeration and the Environment*, London: Greenpeace.
—— (1998a) 'Scotland's West Coast Hit by Dramatic Increase in Sellafield Radioactive Pollution', *Greenpeace Press Release*, 7 July.
—— (1998b) 'Sellafield Pigeons Classified as Nuclear Waste', *Greenpeace Press Release*, 12 March.
Gregg, P., Machin, S. and Szymanski S. (1993) 'The Disappearing Relationship between Directors' Pay and Corporate Performance', *British Journal of Industrial Relations* 31(1): 1–9.
Guardian leader (1998) 'The Grade Gap', *Guardian*, 17 Aug., p. 17.
Haas, P. (1992a) 'Introduction: Epistemic Communities and International Policy Coordination', *International Organisation* 46(1): 1–35.

Haas, P. (1992b) 'Banning Chlorofluorocarbons: Epistemic Community Efforts to Protect Stratospheric Ozone', *International Organisation* 46(1): 187–244.
Habermas, J. (1971) *Toward a Rational Society*, London: Heinemann.
—— (1982) 'A Reply to my Critics', pp. 219–83, in *Habermas: Critical Debates*, ed. Thompson, J. B. and Held, D., London: Macmillan.
—— (1987) *The Theory of Communicative Action*, vols 1 and 2, Cambridge: Polity Press.
Hajer, M. (1997) *The Politics of Environmental Discourse*, Oxford: Oxford University Press.
Hall, P. and Taylor, R. (1996) 'Political Science and the Three New Institutionalisms', *Political Studies* XLIV: 936–57.
Hall, T. (1986) *Nuclear Politics*, Harmondsworth, Middlesex: Penguin.
Handy, C. (1995) 'The Empty Raincoat', Sydney: Arrow Business Books.
Hardin, G. (1968) 'The Tragedy of the Commons', *Science* 162: 1234–48.
Hay, C. (1995) 'Structure and Agency', in Marsh, D. and Stokes, G. (eds), *Theory and Methods in Political Science*, Basingstoke, Hants: Macmillan Press.
—— (1997) 'Divided by a Common Language', *Politics* 17(1): 45–52.
Hay, C. and Watson, M. (1999) 'Neither Here Nor There? New Labour's Third Way Adventism', in Funk, L. (ed.), *The Economics and Politics of the Third Way*, Hamburg: Lit Verlag.
Hay, C. and Wincott, D. (1998) 'Structure, Agency and Historical Institutionalism', *Political Studies* 46(5): 951–7.
Hayward, T. (1994) *An Introduction to Ecological Thought*, Cambridge: Polity Press.
Hede, A. (1991) 'Managerial Reform and Performance – the Case of the Victorian SES', *Australian Journal of Public Administration* 50(4): 490–504.
Heilbroner, R. (1974) *An Inquiry into the Human Prospect*, New York: Norton.
Hillman, M. (1976) 'Social Goals for Transport Policy', in *Proceedings of the Conference of the Institute of Civil Engineers*, in Transport and Society, Institution of Civil Engineers, pp. 13–20.
Howarth, D. (1995) 'Discourse Theory', in Marsh, D. and Stokes, G. (eds), *Theory and Methods in Political Science*, Basingstoke, Hants: Macmillan Press.
Humphreys, D. (1998) 'The International Relations of Global Environmental Change', in Chan, S. and Wiener, J., *International Currents of Thought in the Twentieth Century*, London: I. B. Taurus.
Hutton, W. (1996) *The State We're In*, London: Vintage.
Illich, I. (1973) *Tools for Conviviality*, London: Calder and Boyars.
Jackson, T. (1991) 'Efficiency without Tears, No-Regrets Energy Policy to Combat Climate Change', London: Friends of the Earth.
Jacobs, M. (1997) 'Environmental Valuation: Deliberative Democracy and Public Decision-making Institutions', in Foster J., (ed.), *Valuing Nature*, London: Routledge; cited by Barry, J. (1999) *Rethinking Green Politics*, London: Sage, p. 146.
James, O. (1991) 'Britain On the Couch', London: Century.
Jasper, J. (1990) *Nuclear Politics*, Princeton, NJ: Princeton University Press.
Jones, P., Osborn, T., Wigley, T., Kelly, P. and Santer, B. (1997) 'Comparisons between the Microwave Sounding Unit Temperature Record and the Surface Temperature Record from 1979 to 1996: Real Differences or Potential Discontinuities?' *Journal of Geophysical Research* 102(D25).
Jordan, G. and Maloney, W. (1996) 'How Bumble-bees Fly: Accounting for Public Interest Participation', *Political Studies* XLIV: 668–85.

Judge, K. (1995) 'Income Distribution and Life Expectancy: a Critical Appraisal', *British Medical Journal* 311: 1282–85.
Kaplan, G., Pamuk, E., Lynch, J., Cohen, R. and Balfour, J. (1996) 'Inequality in Income Mortality in the United States: Analysis of Mortality and Potential Pathways', *British Medical Journal* 312: 999–1003.
Karasek, R. and Theorell, T. (1990) *Healthy Work: Stress, Productivity, and the Reconstruction of Working Life*, New York: Basic Books.
Kass, L. (1993) 'Introduction: The Problem of Technology', pp. 1–26, in A. Melzer, J. Weinberger and M. Zinman, *Technology in the Western Political Tradition*, Ithaca, NY: Cornell University Press.
Kawachi, I. and Kennedy, B. (1997) 'Health and Social Cohesion: Why Care about Income Inequality?' *British Medical Journal* 314: 1037–40.
Kennedy, B., Kawachi, I. and Prothrow-Stith, D. (1996) 'Income Distribution and Mortality: Cross Sectional Ecological Study of the Robin Hood Index in the United States', *British Medical Journal* 312: 1004–7.
Komanoff, C. (1982) *Nuclear and Coal Capital Costs, Regulation and Economics*, New York: Van Nostrand Reinhold.
Kompier, M. and Cooper, C. (1999) *Preventing Stress, Improving Productivity*, London: Routledge.
Konig, T. and Brauninger, T. (1998) 'The Formation of Policy Networks, Preferences, Institutions and Actors' Choice of Information and Exchange Relations', *Journal of Theoretical Politics* 10(4): 445–71.
Lazear, P. (1996) *Performance Pay and Productivity*, Cambridge, MA: National Bureau of Economic Research.
Lean, G. (1999) 'The Humbling of a GM Giant', *Independent on Sunday* (3 Oct.), p. 18.
Levi, L. (1972) 'Conditions of Work and Sympathoadrenomedullary Activity: Experimental Manipulations in a Real Life Setting', in *Stress and Distress in Response to Psychosocial Stimuli*, ed. L. Levi, Acta Med Scand Suppl 528(191); cited by R. Karasek and T. Theorell (1990) *Healthy Work: Stress, Productivity, and the Reconstruction of Working Life*, New York: Basic Books, p. 106.
Liberatore, A. (1995) 'The Social Construction of Environmental Problems', pp. 59–84, in P. Glasbergen and A. Blowers, *Environmental Policy in an International Context – Perspectives*.
Licinio, J., Gold, P. and Wong, M. (1995) 'A Molecular Mechanism for Stress Induced Alterations in Susceptibility to Disease', *Lancet* 346(8967): 104–6.
Litfin, K. (1994) *Ozone Discourses*, New York: Columbia University Press.
Liukkonen, P., Cartwright, S. and Cooper, C. (1999) 'Costs and Benefits of Stress Prevention in Organisations: Review and New Methodology', pp. 33–51, in *Preventing Stress, Improving Productivity*, M. Kompier and C. Cooper (eds), London: Routledge.
Lohmann, L. (1991) 'Dismal Green Science', *Ecologist* 21(5): 194–5.
Lovelock, J. (1988) *The Ages of Gaia*, Oxford: Oxford University Press.
Lovins, A. and Lovins, H. (1998) *Climate Protection for Fun and Profit*, Snowmass, CA: Rocky Mountain Institute.
Lowery, C., Petty, M. and Thompson, J. (1995) 'Employee Perceptions of the Effectiveness of a Performance-based Pay Program in a Large Public Utility', *Public Personnel Management* 24(4), pp. 475–92.
Lucas, N. (1985) *Western European Energy Policies*, Oxford: Clarendon Press.

Lukes, S. (1974) *Power: A Radical View*, London: Macmillan.
McCormick, J. (1991) *British Politics and the Environment*, London: Earthscan.
—— (1997) *Acid Earth, the Politics of Acid Pollution*, London: Earthscan.
Mackenbach, J. P. (1994) 'Socioeconomic Inequalities in Health in the Netherlands: Impact of a Five Year Research Programme', *British Medical Journal* 309: 1487–91.
McKerron, G. (1992) 'Nuclear Costs, Why Do They Keep Rising?', *Energy Policy* 20(5) (July).
Macleod, D. (1996) 'Inspectors to Weed out Bad Teachers', *Guardian*, 21 March 1996, p. 10.
Maloney, W. A., Jordan, A. G. and McLaughlin, A. M. (1994) 'Interest Groups and Public Policy: The Insider/Outsider Model Revisited', *Journal of Public Policy* 14: 17–38.
Marmot, M., Bosma, H., Hemingway, H., Brunner, E. and Stansfeld, S. (1997) 'Contribution of Job Control and Other Risk Factors to Social Variations in Coronary Heart Disease Incidence', *Lancet* 350: 235–9.
Marsden, D. and French, S. (1998) 'What a Performance – Performance Related Pay in the Public Services', London School of Economics: Centre for Economic Performance.
Marsh, D. (ed.) (1983) *Capital and Politics in Western Europe*, London: Frank Cass.
Marsh, D. (ed.) (1998) *Comparing Policy Networks*, Buckingham: Open University Press.
Marsh, D., Buller, J., Hay, C., Johnston, J., Kerr, P., McAnulla, S. and Watson, M. (1999) *Postwar British Politics in Perspective*, Cambridge: Polity Press.
Marsh, D. and Rhodes, R. A. W. (1992) *Policy Networks in British Government*, Oxford: Oxford University Press.
Massey, D., Allen, J. and Pile, S. (1999) *City Worlds*, London: Routledge.
Mazey, S. and Richardson, J. (1992) 'Environmental Groups and the EC: Challenges and Opportunities', *Environmental Politics* 1(4): 109–28.
Mellor, M. (1992) 'Green Politics: Ecofeminist, Ecofeminine or Ecomasculine?' *Environmental Politics* 1(2): 229–51.
Melzer, A. (1993) 'The Problem with the Problem of Technology', pp. 287–322, in A. Melzer, J. Weinberger and M. Zinman, *Technology in the Western Political Tradition*, Ithaca, NY: Cornell University Press.
Milne, S. (1998) 'Managers under Stress "New Workplace Bullies"', *Guardian* (1 Dec.), p. 7.
Murray, L. (1998) 'The Effects of Group Structure and Rearing Strategy on Personality in Chimpanzees', *International Zoological Yearbook* 1998, 36: 97–108.
Naess, A. (1973) 'The Shallow and the Deep, Long-range Ecology Movement: A Summary', *Inquiry* 16: 95–100.
Netterstrom, B. (1999) 'Denmark: Self-Rule on Route 166: An Intervention Study among Bus Drivers', pp. 175–94, in *Preventing Stress, Improving Productivity*, M. Kompier and C. Cooper (eds), London: Routledge.
Nordhaus, W. (1994) 'To Slow or Not to Slow: The Economics of the Greenhouse Effect', *Economic Journal* 101(407): 920–37.
North, R. (1995) 'End of the Green Crusade', *New Scientist* (4 March): 38–41.
Norton, B. (1991) *Toward Unity among Environmentalists*, New York: Oxford University Press.
OECD (1993) *Digest of Environmental Data*, Paris: OECD.

Olson, D. (1986) 'Mining the Human Sciences: Some Relations between Hermeneutics and Epistemology', *Interchange* 17(2).
Olson, M. (1965) *The Logic of Collective Action*, Cambridge, MA: Harvard University Press.
Ophuls, W. (1973) 'Leviathan or Oblivion?' in H. Daly (ed.), *Toward a Steady State Economy*, San Francisco: Freeman.
Opp, K. (1986) 'Soft Incentives and Collective Action: Participation in the Anti-Nuclear Movement', *British Journal of Political Science* 16: 87–112.
O'Riordan, T. (1981a) *Environmentalism*, London: Pluto.
—— (1981b) 'Environmentalism and Education', *Journal of Geography in Higher Education* 5(1): 3–18.
O'Riordan, T. and Jordan, A. (1995) 'The Precautionary Principle in Contemporary Environmental Politics', *Environmental Values* 4: 3.
Ostrom, E. (1990) 'Governing the Commons, the Evolution of Institutions for Collective Action', Cambridge: Cambridge University Press.
Pacey, A. (1990) *Technology in World Civilisation*, Oxford: Blackwell.
Paehlke, R. (1989) *Environmentalism and the Future of Progressive Politics*, New Haven: Yale University Press.
Paterson, M. (1996) *Global Warming and Global Politics*, London: Routledge.
Pearce, D., Markandya, A. and Barbier, E. (1989) *Blueprint for a Green Economy*, London: Earthscan.
—— (1993) *Blueprint 3: Measuring Sustainable Development*, London: Earthscan.
Pepper, D. (1993) 'Anthropocentrism, Humanism and Eco-socialism: A Blueprint for the Survival of Ecological Politics', *Environmental Politics* 2(3): 428–52.
Peterson, M. J. (1992) 'Whalers, Cetologists, Environmentalists, and the International Management of Whaling', *International Organisation* 46(1) (Winter): 147–86.
Porritt, J. (1984) *Seeing Green, the Politics of Ecology Explained*, Oxford: Blackwell.
Poster, M. (1984) 'Foucault, Marxism, and History: Mode of Production versus Mode of Information', Cambridge: Polity.
Purnell, S. (1985) 'The Effects of Strategic Network Changes on Traffic Flows', PRA Note 4/BP105, London: Greater London Council, Transportation and Development Department.
Putnam, R. (1993) 'Making Democracy Work: Civic Traditions in Modern Italy', Princeton, NJ: Princeton University Press.
Quine, W. (1972) 'Epistemology Naturalized', in J. Royce and W. Rozeboom (eds), *The Psychology of Knowing*, New York: Gordon and Breach.
Rafferty, F. and Barnard, M. (1998) 'Will it Divide or Motivate? Performance Related Pay', *Times Educational Supplement* (4 Dec.), p. 5.
Remmert, H. (1980) *Ecology – A Textbook*, New York: Springer-Verlag.
Rhodes, R. (1995) *Dark Sun*, New York: Simon & Schuster.
Richardson, J. and Jordan, A. (1979) *Governing under Pressure: British Democracy in a Post-Parliamentary Democracy*, Oxford: Martin Robertson.
Rifkin, J. (1995) *The End of Work*, New York: Putnam's Sons.
Robertson, D. and Symons, J. (1996) 'Do Peer Groups Matter? Peer Group versus Schooling Effects on Academic Attainment', London: Centre for Economic Performance, London School of Economics.
Robertson, J. (1978) *The Sane Alternative*, Oxfordshire: Robertson.
—— (1985) *Future Work*, London: Gower/Maurice Temple Smith.
Robin, J. and Dominguez, J. (1999) *Your Money or Your Life*, London: Penguin.

Robinson, P. (1997) *Literacy, Numeracy and Economic Performance*, London: Centre for Economic Performance, London School of Economics.
Rorty, R. (1980) *Philosophy and the Mirror of Nature*, Oxford: Blackwell.
—— (1982) *Consequences of Pragmatism*, Hemel Hempstead: Harvester Wheatsheaf.
—— (1986) 'Foucault and Epistemology', in D. Cozens Hoy, *Foucault: A Critical Reader*, Oxford: Blackwell, pp. 41–9.
Rosen, S. (1990) 'Contracts and the Market for Executives', National Bureau of Economic Research Discussion Paper No. 3542; cited by Gregg *et al.* (1993).
Ruble, D. (1991) 'Changing Patterns of Comparative Behaviour as Skills are Required: A Functional Model of Self-evaluation', in J. Suls *et al.*, *'Social Comparison: Contemporary Theory and Research'*, N. J.: Lawrence Erlbaum; cited by James (1997: 118).
Rudig, W., Franklin, M. and Bennie, L. (1993) 'Green Blues: The Rise and Decline of the British Green Party', Strathclyde Government and Politics paper No. 95, Glagow: University of Strathclyde Department of Government.
Runciman, W. (1966) *Relative Deprivation and Social Justice*, London: Routledge & Kegan Paul.
Ryle, M. (1988) *Ecology and Socialism*, London: Radius.
Sachs, W. (ed.) (1992) *'The Development Dictionary'; A Guide to Knowledge as Power*, London: Zed.
—— (1993) 'Global Ecology and the Shadow of Development', in W. Sachs, (ed.), *Global Ecology: A New Arena of Political Conflict*, London: Zed Books.
Sale, K. (1984) 'Bioregionalism – A New Way to Treat the Land', *Ecologist* 14(4): 167–73.
Sandler, T. (1992) 'After the Cold War; Secure the Global Commons', *Challenge* 3514, pp. 16–33.
Sapolsky, R. and Share, L. (1994) 'Rank-related Differences in Cardiovascular Function among Wild Baboons – Role of Sensitivity to Glucocorticoids', *American Journal of Primatology* 32(4): 261–75.
Saurin, J. (1993) 'Global Environmental Degradation, Modernity and Environmental Knowledge', *Environmental Politics* 2(4): 46–64.
Saward, M. (1992) 'The Civil Nuclear Network in Britain', in D. Marsh and R. Rhodes, *Policy Networks in British Government*, Oxford: Oxford University Press.
Schor, J. (1992) *The Overworked American*, New York: Basic Books.
—— (1998) *The Overspent American*, New York: Basic Books.
Schumacher, E. (1979) *Good Work*, London: Jonathan Cape.
—— (1993) *Small Is Beautiful*, London: Vintage.
Seaver, B. (1997) 'Stratospheric Ozone Protection: IR Theory and the Montreal Protocol on Substances that Deplete the Ozone Layer', *Environmental Politics* 6(3): 31–67.
Self, P. (1993) *Government by the Market*, London: Macmillan.
Shapiro, M. (1981) *Language and Political Understanding*, New Haven, CT: Yale University Press.
Shiva, V. (1989) *Staying Alive: Women, Ecology and Development*, London: Zed Books.
Simon, J. (1981) *The Ultimate Resource*, Princeton, NJ: Princeton University Press.
Sloep, P. and van Dam-Mieras, M. (1995) 'Science on Environmental Problems', in P. Glasbergen and A. Blowers (eds), *Environmental Problems in an International Context* (Perspectives), London: Arnold.

Smith, M. J. (1993) *Pressure, Power and Policy*, London: Harvester Wheatsheaf.
Smith, P. and Morton, G. (1993) 'Union Exclusion and the Decollectivisation of Industrial Relations in Contemporary Britain', *British Journal of Industrial Relations* 31(1): 97–114.
STE Research (1996) 'Stressing Performance 5, A Report on Performance Pay, Workloads and Stress among Managerial and Professional Employees in BT plc', Teddington, Middlx: Society of Telephone Engineers.
Thomas, R. (1998) 'Politics is Dead. Science is Sexy', *Observer*, 28 Jan., p. 28.
Toke, D. (1995) *The Low Cost Planet*, London: Pluto.
—— (1997) 'Power and Environmental Pressure Groups', *Talking Politics* 9(2): 107–15.
—— (1998) 'The Politics of Green Energy', *Environmental Politics* 7(2): 166–73.
—— (1999a) 'Epistemic Communities and Environmental Groups', *Politics* 19(2): 97–102.
—— (1999b) *Community Wind Power*, Milton Keynes: Network for Alternative Technology and Technological Assessment.
—— (2000) 'Policy Network Creation – The Case of Energy Efficiency', *Public Administration* (forthcoming).
Veroff, J. (1981) *The Inner American – A Self Portrait from 1957–76*, New York: Basic Books; cited by James (1997).
Vincent, A. (1992) *Modern Political Ideologies*, Oxford: Blackwell.
Virgin, C. and Sapolsky, R. (1997) 'Styles of Male Social Behaviour and their Endocrine Correlates among Low Ranking Baboons', *American Journal of Primatology* 42(1): 25–39.
von Storch, H. and Stehr, N. (1997) 'Climate Research: The Case for the Social Sciences', *Ambio* 26(1): 66–71.
Walberg, P., McKee, M., Shkolikov, V. and Leon, D. (1998) 'Economic Change, Crime and Mortality Crisis in Russia: Regional Analysis', *British Medical Journal* 317: 312–18.
Walsh, J. (1993) 'Internalisation versus Decentralisation – an Analysis of Recent Developments in Pay Bargaining', *British Journal of Industrial Relations* 31(3): 409–32.
Ward, H. (1995) 'Rational Choice Theory', in D. Marsh and G. Stoker (eds), *Theory and Methods in Political Science*, Houndmills, Hants: Macmillan.
—— (1996) 'Game Theory and the Politics of Global Warming: The State of Play and Beyond', *Political Studies* XLIV: 850–71.
Wardle, C. (1995) 'NFFO-3 Announcement', Renewable Review Supplement, February 1995.
Watson, M. (1999) 'Globalisation and the Development of the British Political Economy', pp. 125–44, in Marsh, D. *et al.*, *Postwar British Politics in Perspective*, Cambridge: Polity.
Weale, A. (1992) *The New Politics of Pollution*, Manchester: Manchester University Press.
Weart, S. (1988) 'Nuclear Fear', Cambridge, MA: Harvard University Press.
Wentz, F. and Schabel, M. (1998) 'Effects of Orbital Decay on Satellite-derived Lower Troposphere Temperature Trends', Letter to *Nature* 394, 13 Aug.: 661–4.
Wilkinson, R. (1992) 'Income Distribution and Life Expectancy', *British Medical Journal* 304: 165–8.
—— (1995) 'Commentary: A Reply to Ken Judge: Mistaken Criticisms Ignore Overwhelming Evidence', *British Medical Journal* 311: 1285–87.

Wilkinson, R. (1996) *Unhealthy Societies: The Afflictions of Inequality*, London: Routledge.
Witt, R., Clarke, A. and Fielding, N. (1998) 'Crime, Earnings Inequality and Unemployment in England and Wales', *Applied Economic Letters* 5(4): 265–7.
Winter, M. (1996) *Rural Politics*, London: Routledge.
World Commission on Environment and Development (WCED) (1987) *Our Common Future*, Oxford: Oxford University Press.
Worster, D. (1993) 'The Shaky Ground of Sustainability', in W. Sachs (ed.), *Global Ecology: A New Arena of Political Conflict*, London: Zed Books.
Yearley, S. (1990) *The Green Case: A Sociology of Environmental Issues, Arguments and Politics*, London: Harper-Collins.
Zenger, T. (1992) 'Why Do Employers Only Reward Extreme Performance: Examining the Relationships among Performance, Pay, and Turnover', *Administrative Science Quarterly* 37(2): 198–219.

Index

Acheson, D., 94
acid rain, 9, 50, 55
accountants, 79
Addison, P., 83, 127
affluence, 60, affluent society, 75
agents, 2, 10, 11, 17–20, 46, 189–90
agriculture, 175, 179
air pollution, 44
Aitkenhead, D., 121
Alkali Acts, 181
alternative energy, 5
altruism, 89
American Political Science Review, 35
Amnesty International, 40
Anderson, D., 24
Anderson and Leal, 191
Andrews, C., 153
A New Deal for Transport, 22
animals, 164
Animal Rights, 67
Animal Welfare, 67
Antarctic, 39
anthropocentric, 67, 163, 191
anti-nuclear movement, 43, 175
Archaeology of Knowledge, 8, 18
Arkin, A., 109, 146
Arrhenius, 61
Ashby E. and Anderson, M., 60, 181
Association for the Control of Radiation in the West (ACRO), 73
atmospheric nuclear testing, 60
atomic energy, *see* nuclear power
Atomic Energy Acts, 32
Atomic Energy Commission (AEC), 32–33
Australia, 95
Automation, *see* computerisation
Autonomy, 137–8

Bachrach P. and Baratz, M., 18
Bacon, F., 180
Ball, S., 123
Bangladesh, 93

Barry, J., 88, 89, 90, 163
Bates, D., 48
Baumann, Z., 105, 130, 143
Beck, U., 72, 103, 110, 176–9
Bentham, J., 104
Benton, T., 134, 164
Berlin Wall, 76
Beveridge Report, 131
Bhuddism, 141
biospherical egailitarianism, 87
Birmingham, 154
Bishop, J., 95
Black Death, 178
Blair, T., 80
Blaikie, P., 168
Blanchflower D. and Oswald, A., 104
Blix, J. and Heitmeiller, D., 152–4, 192
Blom-Hansen, J., 46
Blowers, A., 177
Blunkett, D., 125
Boardman, B., 13
Bobak, M., 100, 192
Bohemia, 73
Bookchin, M., 147
Boston, 154
botany, 9
Brazil, 73
Brent Spar (oil platform), 73
British Antarctic Survey, 12, 15
British Energy, 28
British Nuclear Fuels, 169
British Telecom, 116
Brooks, G., 121
Brown (Gordon), 113
Brundtland Report, 58, 64, 132
Brynner, J. and Steedman, J., 122
BSE, 182
Buchanan, J., 78
Burchell, B., 110

Caesium, 73, 169
Calculus, 45, 46, 56, 57
Caldeira, T., 130

206 Index

call centres, 109, 116, 141, 145–6
Callendar, G., 61
Calvinist Church, 142
Canada, 95
cancer, 98, 99
capitalism, 7, 79, 111, 136, capitalist interests, 25–6
Capra, F., 133
car emissions, 39
carbon dioxide, 12, 27, 30, 52–6, 61
Carson, R., 59, 61, 182
Carter, J., 15
Carvel, J., 121
Carvi, R., 142
catecholamine, 99, 114
causal beliefs, 65
Center for Environmental Technology, 154
Center for Science and the Environment (India), 70
Central Electricity Generating Board (CEGB), 28
central planning, 91
cerium, 73
cetologists, 66, 67
CFCs, 12, 13, 15–16, 38, 51, 66, 69–70, 72, 73
Chakravorty, R., 13
Cherfas, J., 67
Chernobyl, 167, 177
Chicago, 154
Chile, 159
China, 178
chlorofluorocarbons, see CFCs
Cholera, 178, 180
Christian Church, 180, see also Calvinist and Lutheran Church
Christoff, P., 12
Churchill, Winston, 100
citizenship, 130–2
Clarke P. and Wilson, J., 42
class, 7, 30, 83, 107, 111, 122, 124, 125, 129–31, 139, 185, 187
Clean Air Act (UK), 181
climate change, see global warming
Clinton, B., 11, 80
coal, 30,
cognitive structures, 46
Cole, G., 83

collective action failure, 37, 38
collective action success, 45
commoner, 60, 175, 182
common property resources, 44–5, 47, 56
communism, 76
community banking, 154
competition, (see also perfect competition), 1, 7, 75, 76, 77, 80, 84, 85, 86, 89, 91, 103, 108, 109, 111, 112, 113, 118–26, 130, 149, 159, 185, 186, 187, 188
comprehensive schools, 125
computers, 140, 172
computerisation, 137
Conaty, P., 154–8, 192
conditioning, classical and operant, 166
conservatives, 118, 119
conspicuous spending, 107
constructivism, 65
consumers, 15–16, 104, 105–6, 130–2, 135, 140, 143, 146, 151, 185, 186, 187, 190
consumption, see consumers
consumerism, see consumers
contingent valuation, 53–4
convivial technology, 138, 139
co-operation, 7, 45, 126–8, 134, 141, 143, 186
Copenhagen, 145
Cortisol, 99
Costa Rica, 92, 93, 94
cost-benefit analysis, 34
Council for the Protection of Rural England (CPRE), 22
craftsmen, 179
credit unions, 156
crime, 96
critical realism, 164, 192
critical theorists, 62
Cuba, 93, 94
cultural explanations, 36, 46, 48
Cumbler, J., 181
Curtis R. and Hogan, E., 32
Czechoslovakia, 99–100

Dahl, R., 19
dark green, 174, 176, 192
Darier, E., 6

Daugbjerg, C., 47
Davidson, A., 8
Davies, N., 125
Dawkins, R., 166
decentralisation, 170, 192
decision latitude, 99, 101, 107
decision rules, 46
decommissioning, 28
deep ecology, 87
deforestation, 13
Demeritt, D. and Rothman, D., 53
democracy, 156, 161
Denmark, 141
Denver, Colorado, 118
Department of Education and Employment (DFEE), 118, 119, 120, 122, 123
Department of the Environment, Transport and the Regions (DETR), 22, 31
Department of Trade and Industry (DTI), 27, 31
Department of Transport, 22, 24
depression, 99, 102, 108
Descartes, R., 180
Devall, B. and Sessions, G., 87
Diderot, D., 106, 149
Discourses of the Environment, 6
Discourse theory, 2, 5–35, 160, 163
  discourse coalitions, 5
  dominant discourse, 6, 11, 14, 16, 17, 23, 49, 56, 72, 76, 84
  discursive construction, 111
  discursive formation, 8
  discursive transformation, 11–15
  embedded discourse, 11–16
diversity, 168, 173
Dobson, A., 71, 87, 174
dogs, 165
Doherty, B., 17
Dow Chemicals, 13
Dowding, K., 20, 35
Downs, 36, 38–9, 41
downshifting, 149–54
downsizing, 110, 153, 185
Dryzek, J., 5, 49
Dudley, G. and Richardson, J., 21, 23
Dunlap, R., 12
Du Pont, 15

Durkheim, E., 123

Earth First! 21
Earthwatch, 73
Easter island, 173
Eckersley, R., 67, 87, 134, 191
ecocentric, 67, 87, 192
ecological stewardship, 88, 147
ecologism, 87
ecology, 8–9
Ecology as Politics, 135, 136
economic growth, 75, 76, 86, 112, 128, 142, 159
economics, 77
eco-warriors, 23, 63
education, 107, 118–28, 171
Edwards, R., 169
egalitarian, 125
Ehrlich, P., 60, 175
Elgin, (Dwain), 153
Elliot, D., 27
Elster, J., 36
emissions trading, 50
Employment Service, 116
End of Work, 143
energy efficiency, 13–14, 52–3, 151, 172
engineers, 79, 181
enlightenment, 71, 133, 175, 180
entomology, 60
enunciative modalities, 8, 18, 34
environmental groups, 18, 40, 58, 64–72, 74
Environmental Justice Movement, 186
Environmental Protection Agency, 15
episteme, 170
epistemic communities, 58, 64–72, 73, 74
epistemology, 65, 72, 160, 161–6
equality, 79, 81, 83, 97, 112, 127, 131, 132, 135, 185
equilibrium strategy, 37, 43
European Union (EU), 12, 14, 138
experts, 63, 168, 171, 172
external costs, 78, 86, 91, 114, 117, 123, 126, 130, 132, 183, 185
externalities, *see* external costs
Eyre, N., 53
Ezrahi, Y., 181

Farewell to the Working Class, 136
Farman, J., 170
fat cats, 136
feminism, 149
Ferguson, Sarah, 108
'Fight or flight' response, 98
financial markets, 80
fish, 165
fishing, 39, 44
Fleming, J., 61
forestry, 44
Foucault, M., 5–11, 16, 17–19, 26, 28, 34, 48, 76–7, 113, 136, 160, 161, 173, 174, 185, 187, 188, 189–90
Fowles, R. and Merva, M., 97
Fox, M., 88
Fukuyama, F., 159–60
Funtowicz, S. and Ravetz, J., 169
France, 94, 138
Freedland, J., 63
free rider, 39, 45
Friedman, M., 78, 79
Friends of the Earth, 22, 28, 40, 59
Furedi, F., 111

Gaian, 88
Galbraith, J., 75, 81–2, 97, 108
Galileo Galilei, 180
Gamble, A., 78, 96
game theory, 35, 37, 45, 48, 49, 53–7
Gare, A., 171–3
gas power stations, 30
geneological, 9
Genetically Modified Organisms (GMOs), 14, 63, 64
Germany, 43, 69, 70, 131
Gibbs, L., 186
Giddens, A., 139
Glasbergen, P. and Blowers, A., 191
Gleitman, A., 165
globalisation, 80, 155, 158
global warming, 11, 12–14, 51–6, 61, 71, 172
Goldblatt, D., 138
Goodin, R., 87
Good Work, 155
Gorz, A., 135–41, 143, 146
governmentality, 113, 157
Grade, Lew, 100

grammar schools, 125
Grant, W., 17, 24
Green, D. and Shapiro, I., 35
green globalism, 168
greenhouse gases, 52–6, 69, 169
Greenland, 179
green movement, 59, 62, 131, 133
Green Party (UK), 42–3, 186
  Green parties, 75, 186
Greenpeace, 17, 56, 66, 67, 69, 72, 73, 191
Gregg, P., 114, 115

Haas, P., 64, 65, 66, 68, 69, 70
Habermas, J., 62, 64, 74, 81, 165
Habitat for Humanity, 154
Hajer, M., 5, 15, 59
Hall, T., 7, 26
Hall, P. and Taylor, R., 45, 46
halocarbons, 69
Handy, C., 144
Hardin, G., 35, 37–8, 41, 175
Harwell, 26
Hay, C., 18, 39
Hayek, F., 78
Hayward, T., 87, 132–3
health, 92–101, 167, 173
heart disease, 99–100, 114
Hede, A., 114
hegemony, 7
Heilbronner, R., 38
Heitmeiller, D., *see* Blix
herdsmen, 37–8, 44
hermenuetics, 162
heteronomy, 137–8, 139
Hillman, M., 24
Hiroshima, 60
historical institutionalists, 46, 57
History of Sexuality, 8
holistic, 89, 90, 132–4, 146, 147–9, 176
Hubble telescope, 178
human realism, 164
Hutton, L., 124, 131–2
hydrocarbons, 69, 70
hydrochlorofluorocarbons (HCFCs), 69
hydrofluorocarbons (HFCs), 69
hypertension, 99
Howarth, D., 7

Ice Age, 179
ICI, 74
ideas, 9
identity, 10, 123, 150, 187
ideology, 7
Illich, I., 138, 139–40
India, 70, 73, 156
indigenous peoples, 173
individualisation, 103
individualism, 188
industrialism, 159, 180
industrial revolution, 58, 183
inequality, *see* equality
Inuit, 179
Inland Revenue, 116
Institute of Personnel Development, 108
Intergovernmental Panel on Climate Change (IPCC), 53, 54, 69, 73
International Fund for Animal Welfare, 67
International Organisation, 64, 65, 69
International Whaling Commission, 66
Irish Sea, 72
irrigation, 39, 44
iso-butane, 69
issue attention cycle, 38
Italy, 156

James, O., 102, 106–7, 108, 120
Japan, 66, 95, 141, 147–8
Jasper, J., 34
job insecurity, 110
Jones, P., 170
Jordan, G. and Maloney, W., 40, 42
Judge, K., 97

Kaplan, G., 97
Karasek, K. and Theorell, T., 99, 100, 101, 114, 143
Kass, L., 178
Kawachi, I. and Kennedy, B., 96, 97
Kennedy, B., 96
Kestrel, 165
Keynesian, 7, 76
knowledge (*see also* epistemology), 10–11, 164, 190
Komanoff, C., 33

Kompier, M. and Cooper, C., 144
Kuhn, T., 162
Kyoto Conference, 52, 169

Labour Government, 113, 119–20, 122
Laffer curve, 147
laissez-faire, 113
Lazear, P., 114
lead in petrol, 39
Left, 83–4, 117, 122–3, 124, 125, 136, 187
league tables (schools), 118, 120, 122, 123–6, 128, 185
Learning To Compete, 119
Leopold, A., 87
Leviathan, 44
Levi, L., 114
liberal democracy, 40, 159
Liberal Democrats, 42, 121
Liberatore, A., 70, 167
Licinio, J., 98
life expectancy, 92–3, 94, 174, 185
Liukkonen, P., 144, 148
literacy, 121–2, 127
Litfin, K., 5
local government, 143, 155
Locke, J., 159
London Convention, 12
Los Angeles, 154
Lovelock, J., 88
Lovins, A., 13, 52–3
Lowery, C., 114
Lucas, N., 13
Lukes, S., 19
Lutheran Church, 142

Mackenbach, J., 94
Macleod, D., 118
madness, 6
Maloney, W., 17
Marglin, F., 170
market economy, 76
Marmot, M., 100
Marsden, D. and French, S., 114, 116–17
Marsh, D., 25
Marx, *see* Marxism
marxism, 7, 8, 10, 83, 111, 115, 134–7, 142, 164, 187

materialism, 78, 186
Mauna Loa, 61
Mawhinney, B., 22, 24
Mazey, S. and Richardson, J., 64
McAlpine, 30
McCormick, J., 55, 59, 68
medicine, 8, 18
Mellor, M., 149
Melzer, A., 175
metanarratives, 168
metaphysics, 133, 142
middle classes, see class
middle England, 22
Mills, J., 139
Milne, S., 108
midwest drought, 61
Moby Dick, 66
modernisation, 119, 127
Monsanto, 14
Montreal Convention/Protocol, 12, 68
Morris, W., 142
mortality, 96, 97
Morton, T., 32
Murray, L., 164

Naess, A., 86
Nagasaki, 60
NASA, 170
National Literacy Association, 121
National Power, 28, 30
National Wind Power, 30
neo-classical economics, 53, 188
neo-liberalism, 2, 7, 11, 40, 41, 74, 75–85, 93, 95, 104, 107, 108, 111, 113–14, 115, 116, 125, 126, 128, 129, 131–2, 146, 148, 155, 157–8, 159, 165, 170, 171, 184, 186, 187, 188, 190
neo-liberal discourse, see neo-liberalism
Netherlands, 5, 94, 95
Netterstrom, B., 145
Newbury by-pass, 24
New Economics Foundation, 154
New England, 181
new right, 122
New Road Map Foundation, 153
New York, 93, 152
Nietszche, F., 17, 160
Nihilists, 161

Niskanen, W., 78
Nixon, R., 11
normative theory (see also norms), 1–2, 65, 71, 74, 79, 148, 189
Nordhaus, W., 53
norms (see also normative theory), 47–51, 78, 79, 105
North, R., 71
North Sea, 73
Norton, B., 87
Norway, 66, 95
Nozick, R., 78
Nuclear Electric, 28
nuclear power, 26–34, 137
Nuclear Regulatory Commission (NRC), 33
nuclear weapons, 27, 62, 182
nuclear war, 88, 178
nuclear test ban, 182
nuclear fallout, see radioactivity
numeracy, 121–2, 127

objectivity, 163
OFSTED, 113, 118, 121, 124
oil crisis, 33
Olson (D), 170
Olson (M), 36, 39–41, 77–8, 85
ontology, 162
ownwork, 143
Opp, K., 43
Ophuls, 38
O'Riordan, T., 68, 71, 87
Ostrom, E., 36, 39, 44–7, 49, 50–1, 56, 191
Oxleas Wood, 23
Ozone Discourses, 5
ozone depletion, 5, 11–12, 15–16, 39, 51, 65, 68–70, 71, 167, 170

Pacey, A., 168
Pacific Rim, 119
Paelhke, R., 60, 87
Page, R., 191
part-time work, 137, 143–4
Paterson, M., 56
Paths to Paradise, 138
Pearce, D., 53
permaculture, 175
peer group, 123

Pepper, D., 134
perfect competition, 77–8
performance-related pay (PRP), 2, 11, 76, 80, 109, 112, 113–20, 170, 187
perspectivism, 6, 17
pesticides, 182
Petersen, M., 66, 67
Philadelphia, 154
piecework, 114
Pinochet, A., 159
planned economy, 75, 84, 134, 135
pluralism, 40
political economy, 160, 186
policy community, 28, 46
Politics of Environmental Discourse, 5
Politics of the Earth, 5
Popper, K., 142
population, 60
Porritt, J., 90
positivism, 62, 64, 70–2, 74, 78, 160, 161, 162
 logical positivism, 68, 142
Poster, M., 10
postmodernism, 161, 163, 167, 168–73, 185, 192
power, 5, 9–11, 16, 17 32, 76
precautionary principle, 68
Price-Anderson Act, 32
primary school teachers, 171
Princess Diana, 108
Principles of Political Economy, 139
prisoner's dilemma, 37
privatisation, 191
 privatisation of nuclear power, 28, 30, 31
privileged groups, 40, 51
progress, 62, 108, 125, 127, 159–61, 173–6, 189
psychiatry, 8
psychology, 134, 164–5, 166, 167, 191
psychosocial, 97, 98, 112
public choice theory, 1, 40, 54, 74, 76–81, 83, 85, 149, 188
public ownership, 84, 135
public spending, 82
Purnell, 24
Putnam, R., 96, 156

Quine, W., 170

rabbits, 165
racism, 154
radioactivity, 33, 72, 177, 182
 radioactive fallout, 60, 62
Rafferty, F. and Barnard, M., 113
railways, 135
rationality, communicative and instrumental, 165
rational choice institutionalists, 43, 44–7
rational choice theory (RCT), 1, 34, 35–57, 74, 78, 79, 83, 148, 172, 188–9, 190
Rats, 178
Reagan, R., 15, 81, 90
realism (in international relations), 56
realism (in ontology), 160, 162, *see also* species realism, critical realism, thin realism and human realism
Real World Coalition, 186
reflexive modernisation, 177–8
refrigerators, 70
relative deprivation, 104, 107, 123, 125, 130, 133, 185
relativism, 160, 162, 163
Remmert, H., 9
renewable energy, 26, 29–32, 172
resource conservation, 67
retail therapy, 150
Rhodes, R., 47
Rifkin, J., 156
risk, 176–83
risk society, 103, 110, 177
Roads for Prosperity, 21, 24
road building (United Kingdom), 21–6, 63, 161
Robertson (James), 142–4
Robertson, D. and Symons, J., 122
Robin, J. and Dominguez, J., 152
Robin Hood index, 96
Robinson, P., 127
romantics, romantic movement, 25, 71, 142
Rorty, R., 19, 160, 162–3
Rosen, S., 114, 115
Roseto, Pennsylvania, 96
Rowntree Trust, 110
Rousseau, J., 159

Royal Family, 108
Ruble, D., 120
Rudig, W., 42, 186
Runciman, W., 107
Russell, B., 142
Russia, 94
Ryle, M., 134

Sabatier, P., 42
Sachs, W., 168
Samuelson, R., 52–3
Sandler, T., 51, 54
San Francisco, 154
Sapolsky, R. and Share, L., 102
Saurin, J., 170, 176
Save the Whale, 66
Saward, M., 28
Scandinavia, 55, 141
school league tables, *see* league tables
schools, *see* education
Schor, J., 105, 130, 149
Schumacher, E., 140, 141, 142, 147, 155, 176
scientisation, 62, 63
Scientists, (*see also* epistemic communities), 18, 65, 72, 190
Seattle, 153
Seaver, B., 68
secondary modern schools, 125
secondary teachers, 171
selective incentives, 40
Sellafield, 72–3
Self, P., 76–7, 81
self-employment, 143
self interest, 2, 14, 15–16, 25, 35–7, 41–3, 48, 50, 53, 55, 57, 81–91, 116–17, 124, 131, 141, 148, 183, 187, 188–9
Seratonin, 102
sewerage, 180
Shapiro, I., 7, 35
SHE future, 143
Shell, 73
Shiva, V., 168
short term contracts, 80, 187
Silent Spring, 61
Sloep, P. and Dam-Mieras, M., 191
Sizewell B, 28, 29

small businesses, 155
Smith (Adam), 83, 159, 170
Smith, P. and Morton, G., 115
social construction, 71
social democrats, 81, 134
social green, 91, 134, 146, 149, 154, 156, 157, 185–7
social income, 138
socialism, 83, 134–41
social polarisation, 125, 127, 188
Society of Telephone Engineers (STE), 116
sociologists, 191
soft energy, 52
soft incentives, 41–4
solar power, 13, 26
south, 168
Southampton University, 73
South Shore Investment Bank, 154
sovereign, 190
Spain, 156
species realism, 162–6, 174, 183, 185, 190
Standing Advisory Committee on Trunk Road Assessment (SACTRA), 22
status, 84, 101, 102–5, 107, 108, 109, 116, 123, 128
steam engines, 181
Stockholm Conference, 67
stress, 98–101, 102, 108–112, 114, 116, 117, 122, 126, 128–32, 133, 138, 145–6, 147, 150, 183, 185
stress counselling, 98
stress management, 111, 185
Structures, 20
structuralism, 18
subject, 10–11, 14, 16, 133, 187
subsidiarity, 155
sulphur dioxide, 38
supernatural, 180
surveillance, 48, 109, 174, 190
sustainability, 59, 90
sustainable development, 59
Swampy, 23
Switzerland, 95, 128

tax and taxation, 79, 81, 83, 147, 151, 155, 187

Taylor-Woodrow, 30
Teachers – Meeting the Challenge of Change, 119
techne, 170
technetium, 73
technocentric, 87
technological determinism, 139, 140, 141
technology, 3, 34, 60, 63, 71, 137, 140, 143, 159, 164, 173–83, 190
telephone workers, *see* call-centres
Teller, E., 61
Thatcher, M., 7, 59, 81, 90, 96, 113, 123
The Logic of Collective Action, 39
thin realism, 166
Tibet, 168
tradeable emissions, 191
trade unions, 83, 117, 136, 137, 186
'Tragedy of the Commons', 37
Transport, 5, 20–6
Transport, The Way Forward, 22, 23
Toke, D., 23, 29, 31, 46
transpersonal ecology, 87
truth, 6, 9, 18, 19, 24, 31, 34, 71, 160–7, 174, 183, 184–5, 190
Twyford Down, 23, 24

ultra-violet, 165
underclass, *see* class
unemployment, 139
Union of Concerned Scientists, 33
United Kingdom, 5, 13, 17, 20–32, 26–32, 38, 40, 68, 75, 80, 83, 84, 93, 94, 95, 96, 100, 102, 103, 104, 106, 113,114,115–28, 135, 149, 155, 161, 171, 181, 187
United Nations Conference on Environment and Development, 58
United States, 12, 13, 14, 15–16, 27, 31, 32–4, 38, 68, 75, 84, 86, 91, 93, 94, 95, 96, 97, 98, 102, 104, 105, 113, 114, 115, 118, 147–8, 149, 152–3, 154, 155, 181, 186, 187
universe, 166
university teachers, 171
upscaling, 106
urban regeneration, 154–8

urine, 165
utilitarian, 104

Veroff, J., 102
Ververt monkeys, 103
Vienna Convention, 12
Vincent, A., 86, 88
Virgin, C. and Sapolsky, R., 102
voles, 165
voluntary work, 122, 143
Volvo, 143
Von Storch, H. and Stehr, N., 180

Walberg, P., 94, 97
Walsh, J., 114–15
Ward, H., 35, 48, 54–7
Wardle, C., 30
water (as industry), 135
water pollution, 39
Watson, M., 80–1
wave power, 26
Weale, A., 35
Weart, S., 61, 62
welfare ecology, 67
Wentz, F. and Schabel, M., 170
whaling, whales, 39, 66–8, 71, 168
Whitehall study, 100, 101
Whitelegg, J., 24
Wilkinson, R., 95, 96, 97, 127
wind power, 26, 29
Witt, R., 97
Wittgenstein, L., 18
women, 149
work, 141–6
workers' control, 146
work ethic, 142
working class, *see* class
World Resources Institute, 70
World War II, 26, 60, 61, 83, 111, 127, 155, 182
Wuppman, G., 191
Wynne, B., 169

Yearley, S., 68
Your Money or Your Life, 152, 153

Zenger, T., 114
zoology, 9